Old Man Farming

essays of a rewarded life

by Lynn R. Miller

Old Man Farming
Lynn R. Miller
copyright © 2014 Lynn R. Miller

All rights reserved including those of translation. This book, or parts thereof, may not be reproduced in any form without the written permission of the author or publisher. Neither the author nor the publisher, by publication of this material, ensure to anyone the use of such material against liability of any kind including infringement of any patent. Inquiries should be addressed to the publishers, Davila Art & Books LLC and Small Farmer's Journal Inc.

Publisher
Davila Art & Books LLC
in conjunction with Small Farmer's Journal Inc.
PO Box 1627, Sisters, Oregon 97759
(541)-549-4402
www.lynnrmiller.com

authored by Lynn R. Miller

First Edition October 2014

Library of Congress Catalog Number
ISBN 978-1-885210-25-8

also by Lynn Miller
Why Farm
Farmer Pirates & Dancing Cows
Starting Your Farm
Thought Small (Poetry)
The Glass Horse (Novel)
Ten Acres Enough: The Small Farm Dream is Possible
The Work Horse Handbook
Horses At Work (out of print)
Training Workhorses / Training Teamsters
Horsedrawn Plows & plowing
Haying with Horses
Horsedrawn Tillage Tools
Starting Your Farm
The Mower Book
Complete Barn Book (Out of Print)

cover: The Magnificent Liar, oil on canvas by Lynn R. Miller
used by permission from the private collection of Paul Hunter

dedication

For my father who said to me a dozen years back, "Son, you aren't much of a writer. Time to move on."

So I gave up on writing and decided instead to just put down those things I felt needed saying.

more than biographies of ideas
why not essays as intervention?

*...though I know
it wasn't Heine or Emile Zola I thought
it had to be either Gogol or Dostoyevsky
who threw his arms around the bleeding horse;
and there is so much to say about him I want to
live again so I have time to study him,
for intervening is the only mercy left now,"*

- from *Nietzche* by Gerald Stern

contents

one	what it takes	11
two	nearby	15
three	walking and wondering	23
four	salt lick	33
five	three eggs	39
six	time to farm	47
seven	drumming	57
eight	spinning ladders	65
nine	smaller view	71
ten	cultivating the small farm	79
eleven	spun honey	91
twelve	farming for life	99
thirteen	what I'm looking for is	109
fourteen	to buy a fat pig	119
fifteen	skunks and snails	127
sixteen	fierce plowman	135
seventeen	in the midst	147
eighteen	less talk	155
nineteen	it is who we are	159
twenty	a mulch of time	169
twentyone	charting	183
twenty two	saving for what?	193
twenty three	old man farming	199

premise

We understand the simpler meanings of 'fertility' and 'economics'. As farmers we appreciate that gift of nature which would allow us to set a seed in the soil, grow the plant it creates, and gather its multiplied bounty. The single seed becomes a multitude of seed or fruit or forage. This is a wondrous magic. And the mysteries and vagaries of the host top soil are no less wondrous and magical. Some of us know that we have choices which can result in that top soil's measurable and immeasurable fertility. We can cause it to deplete, maintain, or increase in fertility. The same thing is true, though perhaps more abstract, about all aspects of our farm and farming. There is a general fertility which extends to the reproductive health of our livestock, flows through our pleasure with our efforts, and radiates from the accumulated farm plannings, desires and hopes. It is all about economics and fertility. And it is also all about hard work and the future.

one

what it takes...

Some will remember how it was that Dad never explained, just expected you to know. "No, not that way. To the left, to the left! Haven't you been paying attention?"

Instruction was a ludicrous concept. Water in the nose, fire on the skin, ridicule in the gut, dizzy with pain, nauseous with anxiety, dull with confusion; these were the ways to learn. Those days, for some they may still be today, if you didn't allow yourself to be pulled along you were left behind. And behind was nowhere, no flow, no connection, no justification, no ladders, no doors, no coupon, no pay, no stay, no return.

"Why would I waste myself explaining to a kid or a greenhorn how the thing is done? It's an invitation to questions, the answers to which invite more questions. The work doesn't get done that way. And the kid doesn't learn that way. They either pick it up or they are out of here! Forgiveness and understanding never got the pig cut and wrapped."

I wonder if this 'tough it out' message isn't a main reason why so many of us farmers are fiercely independent?

"Nobody held my hand when I learned to work a team."

Hard to argue with those sorts of valuations of resilience and self-sufficiency. I certainly came from that. But I'll try to argue nonetheless, because today so many are desperate to know what to do. The collective memories of that other great depression frequently suggests that it took toughness and self-reliance to survive when all else failed. But there is also ample evidence of how it was that communities working together made a very great difference in hope and possibility. Or how 'extended' and deep-rooted family held together faith.

Bad days at the bank, sad days on the edges of the river. Millions of good people in this country and others have found themselves in the very depths of economic and emotional depression. It has begun to dawn on those of us who thought ourselves immune, resilient, self reliant, that this minus tide IS taking down ALL boats. No amount of pretending, no amount of analytical gymnastics, hides this terrible fact. But that doesn't stop the opportunistic merchants and priests of denial. Why do I make these observations at this time? What possible good is done by pointing out the painfully obvious? I believe to my core that amidst this depression one of humanity's greatest enemies is alienation, I don't mean as in those protectionist tendencies that alienate countries and cultures from one another (bad enough those), I'm speaking of the close-in alienation that breeds unhealthy suspicions and distance between individuals.

"Keep them away. Their problems aren't ours. We're clean and strong, they're living on the river's edge in a tent. We aren't like that, we're clean and strong."

Wrong. Their problems ARE ours. And you know what? If we could believe that, really act and believe as though we are in this together, their problems would lessen AND the tide would turn. It has nothing to do with commerce, with spending, with government largess. It has EVERYTHING to do with TRUE community. Nothing to do with handouts. Many people know this and act on it daily. The river's edge is peopled not just with the homeless, it is also regularly visited by folks who care and, regardless of their own personal well being, folks who will stop at nothing when it comes to helping those suffering human beings within their ever widening view.

At the same time, within the wider agricultural community, there is a stiffer, longer held tradition of alienation buried deep within the most ornery and tenacious of farming's survivors. Some might take offense at my choice of words and prefer to call them fiercely independent. While they are definitely that, I will add that some of those folks bring upon themselves by choice and consequence a very real alienation from other individuals and community effort. But for this discussion I would point out that this world of farming's independent survivors is a parallel universe which measured against today's depression shows a curious pattern of weaving trajectories. A pattern which may show us what it takes to build acceptable and revitalizing community.

Long after his physical capacities have dwindled to pain and stiffening, what drives the solitary old man to continue bringing in the handful of Guernsey cows to milk? To laboriously split the piles of perfect kindling and stack so meticulously in the perfect woodshed? To struggle with the anxious young horse in tedious repetitious harnessing? To calmly shoot dead the helpless suffering cow? To stoop and pick the wild flowers for his lonely breakfast table? To disassemble the hydraulic pump for the fourth time, carefully replacing the o-ring? To scratch with pencil stub at the scrap of paper, planning a new cross fence he may never be able to build?

After all, this man does not worry about getting a piece of land to farm, he is beyond that. And this man does not worry about family as his wife is dead and gone and his children are far enough away in their own anxieties and longings to be disconnected. And he does not worry about learning how to farm, something that he absorbed in his dirt-fighting youth. Pulling a red hot chunk of steel from the forge fire, or pulling a struggling calf from its young mother's uterus, or pulling hay from the mow, or pulling five dollars from his wallet, he doesn't think much about making his farm pay, that's behind him now. Eighty-five years old, he doesn't think about not having a retirement account or health insurance. If he worries it is about his cows, who will care for them when he wakes up dead? If he worries it is about his land, who will know what to do with this fragile piece of the planet? If he worries it is about his tools and what they evidence. Who will know how to use the device he invented to pull stuck posts out of the ground? Who will know how to spin the drill press head before starting the motor? Who will forgive the old millstone its dips? Who will keep the wooden handles of his screwdrivers oiled? Who is there to honor the craftsmanship he cultivated for nearly a century? For the only honor craftsmanship can use is that which carries forward with the working.

What pushes the lonely old woman to continue the working of her ramshackle ranch? To stitch together, one more time, the tired corner rock crib? To gulp cold coffee after breathing the ammonia-soaked feed dust of the poultry shed? To shoot dead the errant dogs and bury the tortured pieces of dead lambs? To siphon gas from the tractor to put in the pickup truck? To jack up the long heavy gate and balance for one half hour of juggling frustration just to get it lined up to fall back on to its hinge bolts - a job that could have been done in 20 seconds with one additional pair of helping hands?

After all, this woman owns all of this land. She could sell it tomorrow and live worry free for the rest of her days. Her physical aches and pains, her increasing limitations are each and every one met with the internal shrug of unquestioning dedication and ownership. Even so, she'd love to have someone to share the cold morning sunrises with, someone to laugh with and complain about. Someone who never got in the way and had sense enough to keep wood in the fire. Someone who watched and learned without needing to be taught. Otherwise who will honor the craftsmanship she has cultivated for nearly a century? 'For the only honor craftsmanship can use is that which carries forward with the working.'

two

nearby

> *Sometimes we are so close to our efforts that we can't see what's right in front of us, what's close at hand, what's nearby. As small farmers perhaps we should be going to the neighbors at hand and saying 'Use Us.' Allow that people nearby proudly claim that they as a neighborhood are home to your farm. Because 'nearby' ought to be inclusive rather than just a way to see proximity.*

Close at hand, that's what I mean here with nearby. 'Local' along with 'sustainable' are words which have been usurped by advertising and political speak. They have become so all encompassing that they encompass nothing, that they mean little or nothing in the most important discussions of our day. So let's throw them out, scraps for the feral dogs of commerce. Who knows, perhaps these words will come back some day in tight and useful context?

But for now, let's think about some of these fashionable issues of local self-reliance and local foods from the aspect of what constitutes nearby for each of us. Local feels to me to be inviting a formulaic concept for measurement (ie. within 50 miles?) while nearby suggests comfort and culture for ownership and identity. Sure 'local foods' slips off the tongue easily while 'foods from nearby' doesn't have the easy slide. But what I'm suggesting here, more than the words we use, is that the words we picture when we think about concepts and motivations help us to understand what we're working towards.

Speaking of pictures...

Back in the early seventies I farmed in Junction City, Oregon and happened on to a 'subscription-like' relationship with a small close-knit group of customers who were mothered by a gentle and spirited woman I have called Mrs. Jane. (See *'Mrs. Jane's Beans'* in *'Farmer Pirates and Dancing Cows'*). I like to think this was the occasion for my first gut-felt understanding of what *nearby* could mean. For in very short order I came to feel elderly Mrs. Jane nearby in the most comforting sense.

She cared intensely for what was happening on my little farm. She identified with my working rhythms and relationships. Big old chestnut Bud (a Belgian gelding) was her 'boyfriend' and old 'Bobbie' her soul mate. She saw and felt the rainy season for the sticky mud and the slippery oily sheen of pungent fresh manures. Once she took my hand and turned it face up to run her old fingers over my calluses while a sad sweet murmur of a whistle passed her lips. As I've written before, she came to see my farm as her farm. This magical relationship started with a classified ad I had placed months before.

I was trying to figure out just how much of what to plant the next spring and thought to place an ad in the local newspaper which said;
"Organic Vegetables: commit to certain vegetables this summer in set quantities with a small deposit and receive a big discount."

The idea was to have some certainty to guide my planting.

Mrs. Jane responded with very specific needs which quickly developed into the richness of a possessive yet respectful neighborliness. Forty years later I still hold freshest memories of the enchantment that relationship brought to my farming adventure. We were, to each other, nearby and close at hand. We were to each other an abiding comfort and a reassurance. All I offered was an open access to how some of her food was produced. What she got was sustenance in full-color. All she offered was the most receptive of gratitudes. What I got was a lateral fertility and enchantment as a neighborhood cloak.

Close at hand; how it is that a long life might build in a person, around a person, intangibles such as the list of allowances and comforts that carry the 'weight' of an attitude of creativity, enchantment, and fertility.

The enchantment and comfort of "nearby" must be held and protected lest it become ordinary and taken for granted. When this happens, when gratitude is replaced by dullness, the individual melts into a less than necessary placeholder.

How about the 'nearbys" of over there?

I take some risk at confusing the discussion when I say that local, nearby, close at hand, all offer health and character to people outside of the neighborhood of origin. The 'local' foods of Abiqui, New Mexico have a character and an intrinsic power to charm and feed us 'over here' far better than any industrial processed foods; but to someone in Rochester, New York, wonderful as they may be there ain't nothing 'nearby' about New Mexican foods. Yet those imported local foods, local from somewhere else, are worth inclusion if only for the spice of life.

For some that might fly in the face of the theory that eating local also means while you are improving your health with the mysteries of indigenous foods eaten in season you are also adding to your neighborhood's economy. All true. But perhaps we need to reinvent the Marco Polo notion of spice trade on an agrarian-post-neo-modern scale? I like the dream-state picture of a dozen Richard Farnsworths riding lawnmowers across statelines to return with basket loads of exotic foods not otherwise available in their own areas. Little tents propped up in Church parking lots where you have an opportunity to trade four quarts of your own canned tomatoes for a tissue-wrapped pineapple. I still say that "all foods ought to be local" and that there ain't no only way to do anything.

But for the purists in our midst, I say "local" is not just a question of YOUR delineated region. I like to expand the notion of "local" to include "localized", to include "regio-honore" (some might call this "terroire"). It goes well beyond any concern for protecting environment and endangered species, it goes well beyond notions of supporting individual effort and neighborhood, it goes straight to the heart of expanding identity, culture, and village-sufficiency. I say what we are after is a ponderence of food, fiber and shelter which springs from "nearby". When we support our nearby producers we provide incentive for the growth of local self-sufficiencies AND blossoming local culture. (And adding in the handmade notions of a spice trade means cultural comes to embrace affordable elegance.)

A sign in a super-market defined "local" foods as anything deliverable in a day's drive or less, that would suggest 800 miles; for us that means from southern BC to So-California from the Pacific Ocean to Utah. I would argue 'that ain't local', that ain't nearby. That said, I am reminded of the magic I felt when years ago Californian Bob Puls spoke of taking a rented truck to a Nebraska small town church parking lot with a load of his own oranges to trade to those folks. Handmade spice trade.

Nearby, as in of-a-shared-environ: there are places in the world where the divider between near and far is not a gradual overlapping but a hard line - 'through that gate - across that river' - all is changed, the enchantments we hold dear and defining are lost, or worse, seen as the enemy.

Nearby and related as in neighborhood and community.

Finite versus the infinite. Nearby versus who-knows-where-from. Today I read in mainstream media that organic farming, as local, artisanal and hands-on has been outstripped by demand. More people want the stuff than is currently being supplied. So we hear a new form of the old war cry: "get real about the best efficiencies of scale" - that larger industrial-scale organics are the only solution. Mom and pop organic operations can't feed the world - the spurious argument goes. Interesting how these disputations borrow from notions of the finite, a real contradiction in terms.

Finite; as in with limits - and those limits for our discussion not unlike, anatomically speaking, the serosa or peritoneum, that thin single or double walled membrane that serves as a protective outer wall for the abdomen. Is there a peritoneum-like protective delineation, socially speaking, for the abdomen of a region?

So today in the marketplace 'local' is stretched to accomodate notions of attainable. "Can't grow enough around here to feed all these folks so let's push back the boundaries of 'around here' til we have a best chance for success. A kind of gerry-mandering of the definition of local. Set up the district so we can be assured of electing 'our' candidate. Set-up local so that we have a chance of an attainable local self-sufficiency by some finite measure. But that doesn't always work, I might offer that it never works, because some 'places' got no room to push back in to - and from that you get the inevitable paradoxes.

Take Greece or Haiti for example: For differing and nearly absolute reasons these countries are finding that economic and political realities are separating them from the insidious artifice of the rest of the world, an artifice which screams ***the big lie*** -

> "play by the rules of corporate governance and global banking and everything will be just fine".

Well, both of them would like nothing better than to be back in the global tent BUT circumstances will not allow that for the foreseeable future. With both countries, and for different reasons, these are the toughest of times. So to cope they find themselves having to consider a return to their agricultural roots. (Is that so bad?)

Big banking is the problem along with corporate ethos.

Paradox: We have 7 billion people on the planet, if any significant percentage of those wish to eat organic food (instead of industrial) then they (corporate gatekeepers) say we need to ramp-up a targeted industrial by finding ways to industrialize organic farming. Bull-pucker I say, all the pertinent points are being dismissed. It is as if to say it all boils down to tonnage of commodities, that any discussion of cultural aspect is a dangerous diversion from the task at hand - feeding people. Biological notions be damned.

Feeding people must be, at its core, cultural aspect! Otherwise how are we different from cage-raised chickens? We are different because a cage-raised chicken has a 'market' value where as many of the world's poor have no market value. We are different because many of those people who need to be fed can and should be farming, should be raising their own food and food for others. They should be given the chance to be a vital piece of the assurance of their own neighborhood. They should be recognized as of the highest value to the rest of humanity. But that flies in the face of the mathematical shadow-boxing we know as economics.

What we are being told repeatedly is that our purpose as human beings must be to assure a healthy industrial/military/banking complex. That dignified healthy life itself comes only as a possible secondary result of strong corporate markets. Hasn't it become obvious, through disease, poverty, hunger, alienation, war and social cancers that life is not about the ongoing justification of industrial process and corporate profit?

I've read almost all of it, heard more than I wanted to and I'll say it again, it's a big lie this notion (and it's only a political and merchandising notion bolstered by contrived and manipulated statistics) that the only way we can feed the world is through chemically intensive, bio-engineered, maxi-processed vertically-stacked industrial agribusiness. NOT TRUE! Unfettered industrial agribuisness is killing us and the planet!

Give farming back to natural processes, to human beings, to living soils, to embracing and embraceable weather and you feed EVERYONE well - and with increasing soil fertility. It is impossible to say that with industrial agribusiness. IMPOSSIBLE!

What is the seed of the mayhem we feel from industrial agriculture? Big banking is the problem along with corporate ethos.

We praise those wealthy jackals whose entire purpose in life is to manipulate pieces of absentee-ownership so that their stock values sky-rocket. These asymmetric monsters are more to blame for the board room ethic than any other entity. They are feared, revered, loathed, and embraced. Most people can only dream of being like them. For it is in those portfolioed rich that we see limitless wealth and power, idleness and self lubrication. Their bread and butter stems from corporate stock and the shadow boxing world of abstracted instruments of trade. While most of us abhor the corporate ethic which continues to stomp down on the natural world and the true life-giving spirituality of the biologic, we flit around the edges of it all worried that it might change.

We hate them,
we want to be just like them,
we know things must change,
and we fear losing
this crippling socio-economic system.

Dark tragedy with a sprinkling of ridiculous humor.

"Can't we find a way to keep this system while making it more equitable and reversing the planetary cancers?" you ask. I don't know. I can't see down that tunnel. After a lifetime of steadfast belief in the healing and ecologically redemptive power of agrarianism, of a small farm earth, I fail to see how Microsoft, Walmart and Monsanto offer any dignity or hope to the hundreds of millions of people living in hunger and poverty all across the globe. Yet it is always easy to see how even just one successful small farmer offers dignity and hope to a neighborhood and beyond. I know what can happen when that one small farmer is truly seen by others of similar inclination.

As small farmers we need to see each other as nearby, as close at hand. We need to always work to support one another. We need to ask ourselves every day what we might do to save the world. A bit much? I don't think so. I think it might be the song we've been looking for. I think it just might be the song that would keep us at the good work. I think it's the answer and the reason.

three

walking and wondering

Eighty acres of hayfield/pasture isn't much. It lays out ahead, and on either side, in a way that suggests no secrets. (Of course that's not true.) If you can see its edges and feel an intimacy for it, it ain't much or perhaps better put, it's just enough. Me and my old stock dog Lucky waggle along looking out and looking down always surprised to find compressed stories in that green grassy legume expanse. Here it is thick and dark and lush, shading its own piece of fertility to encourage more. Hiding the valuable truth. Over there thin and dry and crusted to suggest a flat rock size of a pickup truck just under the surface and threatening anything that thinks of taking root. And a dandelion and sour dock patch adjoins announcing a confused spot of acid imbalance. Right there Lucky nervously jams his nose down a sage rat hole puffing so hard that his back end does a little hop as his front end digs at the edge. Me, I watch a red tail hawk a distance away sitting atop a juniper fence post. The hawk seeming to say "you're only here because you're no threat to my food supply". And just to my left are the large droppings of elk from the night before. All of it together a gradually undulating carpet of soil entanglements and plant futures.

Old Lucky moves along sure except that every third step seems to hesitate then hiccup his back end and push him sideways. I smile only until I realize my own gait shows similar irregularities. Me, I have to work at having a true step and hiding those inevitable moments when my signal never reaches the offending leg until a jig step, more burp than dance, escapes. Two old dogs that's us. Have to wonder which of us will go first. Me - I've got the odds favoring a decade or more ahead, him - he's got maybe a year or so, but where it counts, out walking the field, we're pretty much evened up. These days, both of us have felt our appetites constrict, both of us have paid a whole lot more attention to our bottom sides than ever before. And our past tolerance for sympathy has easily morphed to a real need.

Lucky and I own this ground. Me 'cuz my name has been on the papers. He because his entire life has been spent right here delineating and encircling. If you connected all his dog urine moments in a dotted line, these last eleven years would amount to three times around the eighty acres or four and a half miles. I know the distance by patch jobs in the barb wire fence. We aren't the only owners. There is the entire rest of the family and all the other animals living here. There is that cosmic history of the place. There are the ghosts of many a grateful Paiute brave and the ever-changing cavalcade of guvmint boys out to prove that permission needs be granted in this neighborhood. Pretend deference or saddle up and move on.

So we keep our heads down and smile most times. It's the same way when the talk swirls around to industrial farming. Keep your head down and smile, otherwise you're pegged as a trouble-maker and a loony. But Lucky and I know some things, and one thing is that the loonies will be inheriting the world. We also know the only way to feed billions of people is if millions of families, ordinary families, work small plots of ground with their hands, minds and hearts. That way the water, soil, seed, and livestock are protected while food is made available. That's the valuable truth - that little people protect water, soil, seed and livestock. And you got to have all four if you're gonna feed the world. Actually that list includes five - because those little people are a mighty important piece of the puzzle. It's a beautiful thing; small farmers with the spirit of gardeners improving air, water, soil and genetic diversity all while feeding their share of the world. Wow. A neat and repeatable solution. But one which worries some people.

These days there's no end to folks who believe that the world is being destroyed by you and me and her and him - doesn't really seem to matter what we're about - there's some official or learned one out there who says you're to blame. And moving up to the top of the world's villainy list right now are farmers! Yep, hard to believe but good scientists and social engineers have come to the ridiculous conclusion that we farmers all of us pollute and abuse even more than factories, chemicals, cities, cars and wars. They've got the statistics to prove it with charts and graphs and everything. And most of them come to the same conclusion; the solution oddly enough, paradoxically enough, must come from more sophisticated industrial agriculture with subsidized research and development into new zowies in bioengineering and chemistry. They say we who farm the old ways with our hands, hearts and minds are destroying the world while factory farms - though they may be 'somewhat' culpable in the overall picture - offer the best solution to feed more people and save the planet. Even my old stock dog thinks this stinks like pig manure in a hair salon.

The factory farms are ONE of the problems. Small farmers are most of the solution. Whoa they say, we farmers use up the water, the electricity, the beneficial carbons, the best land base. Whoa I say - you do everyone a disservice when you lump us small farmers in with every entity invested in food production, processing and distribution.

To say we are small farmers is to say something very important. We are not miners. We are stewards. We are not users. We are husbands. We practice farming methods which retain water and build soils. We embrace low impact approaches to working because of the smaller 'footprint' but also because it suits our economy. We don't poison. We refresh. We harvest with hand and eye. And we distribute the same way. We walk our fields and gardens and 'look' at them and into them because we want to know them. And we want to know that land - because from the knowledge come the right answers to problems and opportunities. We are not factory workers. We are shepherds. We are not tacticians or economists or efficiency experts. We are parents, lovers, artists, and gardeners. We are not landscape architects. We are the landscape. We are not theologians. We are the religion. We are not destroying the planet, we are healing her. We are small farmers.

And we are at war with those who care more for money and power than they do for life. The cleverest of industrial forces understand that in the near future the single arena on this planet which offers the greatest opportunity for profit and power is food - nothing compares not petroleum not gold not armaments not currencies not housing not education not religion not energy not electronics not artificial intelligence - not even Facebook or Google or iTunes. As we move towards 9 billion people who will need to eat every day, the magnetic equation is easy to see.

If we accept the fact that the food supply is already inadequate and industrial agriculture ill-equipped to supply that increase, what then? Supply and demand. Only in this case the demand is tied to life and death urgencies. Think of it, an arena of endeavor upon which literally every human being is dependent. It's not like whether to buy a magazine subscription or a new suit of clothes. You've got to eat. So if every human being on the planet is waiting with a bowl in one hand and coins in the other can't you see what that attracts?

Organized crime has moved from drugs and prostitution to politics/news and now to wholesale identity theft and income tax fraud. They use computers to bilk billions out of the IRS. It's easy, it's relatively pain and risk free. It makes criminals billions of dollars a year. And it has given the mobsters something they never had, a new respect for education. They need more thugs trained to work computers. It's a monster which cries out to be fed new opportunity. This amazing new pool of human excrement is growing by leaps and bounds and is well-positioned to move into the arena of food - yes food. Because the production models, the futures markets, the distribution matrix are all dependent on computers, access to which is wide open to the unscrupulous. It's not a new thing, industrial agribusiness has long been a happy home for criminal minds but, outside of a couple of villains in mega corporations, most of these guys and gals are petty thieves.

Imagine with me what happens when the federal agencies, corporate board rooms and food distribution infrastructure are infiltrated by Harvard-educated gun-toting computer-stroking mafiosos trained to skim a lot more than cream off the food supply. Bizarre you say? Paranoid you think? Well think a little harder. If there is an opportunity to corner a trillion dollars worth of profit from feeding people, who do you suppose is going to show up for their piece of the action? The good guys?

One of our little secrets, something that keeps us hopeful - you can't control the food system IF it is tied directly and largely to millions of independent small farms. Too many cats to herd. But if there are five or six multinational corporations which control food production and distribution, its easy to imagine tying up that bundle into something highly lucrative and larcenous. Don't think for one minute it isn't happening right now.

Speaking of tying up bundles; as we cross the field I see our peafowl in a gaggle wandering the harvested hayfield where it is easier to see the tasty bugs in the stubble. And of course, out in the open as they are, it is easier for the coyotes to see the pea fowl. When our Great Pyrennes guardian dog, Hal, passed on, the coyotes began to move in. Nothing a young coyote likes better than a fresh unplucked peafowl. So they lie in wait, in whatever depression or tufting of grass and weeds gives them cover; lie in wait, perfectly still and incredibly patient, until a peafowl is dumb enough to walk by, then the wild canines bounce, usually from behind. If its a male in summer that means long fancy tail feathers. When the coyote gets a hold of those feathers and the bird flees, all the feathers come out and scatter across the field for us to pick up. (But if its a female with their short tails and the coyotes grab holt - curtains.) We started the winter with twenty-seven birds - free range. Today we have a dozen and only three of those are females. We have friends who are sure to remark upon reading these words, "good riddance those birds are noisy and messy". But I'll let you in on a golden secret if you promise not to tell the others; one male peacock will produce a magnificent tail of over a hundred patterned and luminous feathers. Those feathers are worth a dollar a piece. They regrow those feathers each year. That means nine peacocks can generate one thousand dollars a year. They require NO purchased inputs raised as we do, free-range. Peafowl just may be the most profitable livestock a rancher can raise. And I've figured out how to make the easy wholesale transactions, just take those feathers to flower shops, tatoo parlors and hair salons. Only thing I can't figure out is where to sell the coyote hides, unless the tatoo parlors might...?

I see the peafowl story and our experience with the birds as an analogy for just how exotic diversification might be for the small farmer. (And diversification is still the ticket to allow the small independent farm operator a real chance to succeed.) If you have an attractive nuisance on your farm its in your best interest to carefully analyze how that nuisance might be converted to cash. We have lots of rocks, some of very interesting and even pleasing shape and color. They are a nuisance to our farming. A man once drove up and offered to buy the flat thin rocks for patio paving stones. We live in the high desert and much of our grazing land is peppered by Juniper trees, considered by many a weed and a nuisance, yet every other year or so a nursery business from the city shows up to harvest the berry-ladened limbs for holiday trim. We get paid for that nuisance. I need to find out if those berries are the same what make gin. We have a large herd of elk that come in most nights to take what they want and go. I've been approached by hunters who will pay me for the right to 'harvest' those elk.

Of course most farms are not "encumbered" by all the wild flora and fauna that live with us. But there are some, and quite a few of those ranches, with problems similar to ours, have taken to ranch tourism; that's where folks pay to stay and partake of the outfit's day to day rituals. Hmmm, diversification.

The real point in all of this is that this is how my mind works when the dog and I are walking the field, it wanders with us and then sets off on its own. And that's a very good thing, it's one of the regularly available rewards of this farming life.

Lucky's getting tired. He's hugging the side of my knee as I amble along. Looking down at him I notice a pair of sidecutters in the grass which must have fallen from my pocket on the previous trip across the field. I stop and reach for them and he thinks I'm reaching for his head to scratch it. We both chuckle, the old farmer and the old stock dog. He does it by blowing out, I do it by inhaling. Similar wheezing noise. And I'm fond of these small moments because they have a scale and an immediacy that color days less grey and more vibrant.

That, after all, is what we've been trying to do with this magazine (Small Farmer's Journal) for almost 4 decades now, gather up small moments of rare honesty and vibrance and allow you to discover them in your lap or at the kitchen table. Little pieces of how we truly fit in this our chosen work-a-day world. And the physical format of this magazine as a print edition, ink on paper, has lent itself to this discovery process. Kind of goes together with the image of this old rancher wandering one of his hayfields with an old stock dog friend. Easy to feel the scale and relevance of it. Easy to see, in your mind's eye, the speed of it.

But then today is another time. I think, even more than a hundred years ago when the internal combustion engine offered cars and trucks to replace horses and mules in harness, today the electronic world of the internet is changing how many people walk and wonder each day. Some of us holdouts still want the paper and ink just as we enjoy teams in harness, heirloom seed varieties, a good and trustworthy saddle horse, and flower boxes round the porch. Some of us don't have the patience for any of that, we want what we want at exactly the moment we want it or we go elsewheres and right away. When I was growing up, that was a definition of spoiled.

The hallmark of the spoiled person was and is impatience. Fewer people walk and wonder, today many more sit in one place, fidgeting, leaving the speed to computerized vehicles and virtual realities, demanding to be satisfied RIGHT NOW.

Perhaps measurements of good and bad, right and wrong, appropriate and inappropriate don't always rule the day. For far too long, I as editor of a print magazine, denied that people were changing how they wanted their information; what the information was to include, how it was presented, and how it was delivered. Fossil that I am I honestly believed there would always be a place for paper and ink, that it represented more than the thing whether book or newspaper or magazine - it represented the experience of deep reading, the experience of a personally chosen context for 'seeing' information, a way that the 'thing' seemed to give the reader a sense of ownership and membership all at the same time. The way we hold altogether firmer to the book as we near her ending, not unlike how I seem to pay much closer attention to my old dog Lucky.

But the paper and ink are different from living creatures because they allow that we might give them an extended shelf life. We might store them for future reference. And enjoy their accumulation as if some measure of places we've been or intend to go. Electronic files, digital archives, though attractive in an antiseptic hospitalized sort of way never offer us those aspects. We can't even be assured of their safe-keeping. Sure, books can go up in flames or slowly crumble with age. But they die in our arms. While electronic files can disappear in the press of the wrong key, they can go fugitive and morph into weird cousins of themselves. They can do the supercede waltz and be twirled by the search engines until they drop off the bottom edge into a cyber dumpster. And there has yet to be any wake or funeral for those cyber slough-offs. There is nothing to hold in your arms at the end.

It is after all the end of the copiest. That human expertise married to process which either by hand or machine gives us another dozen hard copies of the Iliad to pass forward. If I were to tell you that Moby Dick is a thousand page novel and that I know this because I have a hard copy and you were to tell me that 'NO, Moby Dick is actually a 58 page novel because the page counter in the PDF archive tool bar so indicates'. Who would you believe? It may matter now but in the near future it may be an absurd question. Because the cliff note version may be all that's saved on the internet.

Our friend Cary Fowler is attempting to store a few pounds of every known seed variety on Earth, all in a Norweigian arctic snow cave. I have been told he already has varieties in storage which the internet either refuses to acknowledge ever existed or boldy and gladly claims are extinct. There are scientists and computer worshippers who put the highest priority on the accuracy of the computer data base universe. They are more concerned that the information is right in cyber space. They do not see the incredible monumental accomplishment of Fowler's Cave and what a gift it is to humanity at this very moment. Don't ask them to choose which to hold dear, please don't ask them.

If you ask me, the answer is no contest, the seeds are far and away more important to our survival. Electronica for electronica's sake is the Tesla dream gone nightmare. But, not to be seen as disparaging 'our' most important inventor, Nicola Tesla, if he were here he would want to remind us that tools should always be seen and treated as just that and no more; tools. Perhaps there are vestiges of Electronica and the Internet which 'might' serve as useful tools in the incredibly important work necessary to save the biological and sociological diversity of mothership Earth? Perhaps, but what be the value of the 'saving' if access is controlled by 'guvmint' and/or corporations?

Pompous as it may sound, I have long seen our publication adventure, *Small Farmer's Journal,* as a form of democratic existential seed-saving. With each issue and the extended community of readers, we have been saving the ideas (read seeds) and methods (again read seeds) and recorded life adventures (read evidence of seeds) that give us our past and future ways of working. A four dimensional multi-generational recipe book for good farming.

Example: The grain binder in its ultimate ground-drive manifestation was (and is) a superlative engineering fete allowing one person to cut, gather, and tie grain into bundles which could be gathered in groups of six or less and deposited in exact spots across the field. This machine featured a complex system of interconnected moving performances from the mowing to the conveying to the gathering to the knotting to the field deposit all requiring timing and precision; and the operator needing to learn how to rub her belly while patting her head - a reference to what it takes to fly a helicopter. Sitting on the binder perch, so much of what happens with this machine seems counter-intuitive. Brilliant engineering, presupposing that farmers would be able to grasp what was required and make the beauty work, and work it did to astounding result. Yet this pivotal engineering moment lasted for just a couple of decades before combines replaced binding/threshing. Should the machine and its concept have been trashed then? No.

I believe we are most fortunate that the Amish farmers and living history farm re-enactments have kept alive all that goes into binder familiarity, we are fortunate that this very specific way of getting a field harvest job done is still, albeit limited, alive for the near future.

But with a tip of the hat to Tesla, we might at this point add that there is some merit to making sure that this information, in a useful and accurate context, is also 'stored' electronically. Because ultimately, tool that it is, the internet still depends on people deciding what information is 'worth' saving and in what form; for example should it be an abbreviated reference to an obsolete piece of machinery or an indepth and complete archive of all that we know? I say it must be the latter or we've lost the 'germination' potential should we need to revisit this technology and methodology again in earnest. Because already we are seeing how arbitrary and fickle people can be with what gets stored on the internet. Complete information in a useful and appropriate context will be critically important. It's a case of where the 'qualitative' may be more encompassing than the 'quantitative'.

And so we come to this question; what is the importance of the qualitative? And why is 'all' better than 'complete' or vice versa?

As Lucky and I continue our walk across the recently mowed hay field pasture we come to the elk trail, a distinct depression wandering a quarter mile from the irrigation lagoon to the corner of the fence. I recall while mowing that, even though I knew it was there I was surprised as the mower dipped and revealed with the cut this wildlife highway, an eighteen inch wide depression in the height of the field. Before the hay was mowed you wouldn't have seen it, it was covered or shaded by the tall grass, you could only 'feel' it when your feet hit solider ground, slightly dipped. But it's there, reinforced by the continuing visits of the herd of elk and occasionally used by the cattle and deer. It connects as it traverses. Just like that binder in our working history of farming. Just like Cary Fowler's seed cave. Just like this old rancher and his limping stock dog.

While much of this new digital age may prove to tax our notions of biodegradeable, as farmers we know the importance of completing cycles. Me and my old dog will sometime add to the fertility in direct ways. We must be careful that this new so-called freedom we feel from the internet be no less tied to the cycles of life.

Freedom is nothing unless it is biodegradable. Knowledge is nothing without qualitative context. Biology will wait out our follies and rule the day once again. Who knows, perhaps the great storms of the future will rearrange our garbage and presumptions just enough for us to actually be intriguing to the next 'superior' species to come along. 'Or not', as the comedians are fond of burping.

Meanwhile I'm planning on continuing to walk my fields with the dog and allowing myself to wonder. We set quite a few young ships to sail through the rippling waters of right livelihood and it feels good to know that. Let's you and I not stop doing that until we are no longer able. Let's find the elegant hand-off.

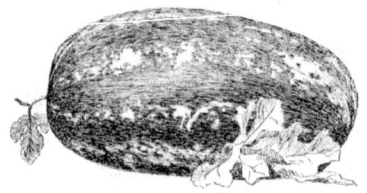

four

"No man is worth his salt who is not ready at all times to risk his well-being, to risk his body, to risk his life, in a great cause." - Theodore Roosevelt

the great salt lick adventure

What follows by way of preface is fiction.

Imagine with me: The mottled and arguing old Datsun pickup slinks alongside the back boundary fence its driver peering intently for telltale browned forest-floor culdesacs the center of which might contain great treasure. Might. Usually in the vicinity of a water source, often near gating, he searches for odd rumpled shapes he appreciates solely for their monetary value. He cannot see the higher value some would place on these ranching artifacts, a higher value born of a marriage of humor, practicality, aesthetics, and exclusivity. Let's call this fictitious salt lick rustler Grover, just 'cuz the name sort of fits. He's not a bad man. He's a natural born comedian, one that instinctively appreciates moments of obscure cultural opportunity - moments when you can just about hear the folks saying "come on, join us, take this curious little community adventure in new directions, make it yours, all bets are off". And prize money to boot! Fiction, remember, just something imagined from a thread of evidence.

Imagine with me again: On the feedstore bulletin board a handmade poster with a picture of a much-licked salt block. The poster asks "Lost salt lick. Has anybody seen my trace mineral salt block (last photographed August 29)? Disappeared between 4 and 5 pm on August 30 from near my water tank at the southeast corner of the summer pasture. Been turning it over every day so I knows when it 'disappeared'. Any information leading to its discovery will be rewarded. I'll give three new blocks for it if I can get it back - no questions asked. If it turns up at the Great Salt Lick you'll be in big trouble! You know who you are. I want my block back, you bum."

Imagine again with me that inside an old abandoned garage out on the edge of Baker City a woman who might be called Sheila operates a pressure washer, working to create new curvilinear indentural shapes into a fresh unlicked livestock salt block, to make of the perfect cube a bubbly lump of surprises. She giggles thinking to herself "those yahoos will never be able to figure out that this one is a fraud, looks like three Charolais bulls stared each other down for a month as they licked on it." Sheila's actually drawn to do this for the cause. Helping in a small way to fund Parkinson's disease research is something she gladly risks her artistic reputation on as she gleefully forges another counterfeit salt lick.

What's this about? What follows is fact.

I've been ranching now for over forty years, that's a good long while and it has solidified for me that the more things change the more they stay the same. Unless some rancher decides… When I first studied bovine genetics with ABS in the early seventies, the shape of cattle ranching seemed to be well set in its modernist form what with artificial insemination, embryo transplants, terminal crosses, advanced rangeland management practices, yield and cutability targets, calving ease, and mineral requirements. On the horizon would come those appropriate technology additions that would alter hand-tools, fencing systems and take us from stock trucks and slide-in pickup racks to the now ubiquitous gooseneck stock trailers. The cowboy stuff, though reinforced by regular revisits, was off in the romantic periphery of the business of cattle raising. If you were looking to get a ranch-hand job you had a better chance if you knew how to run haying equipment, nitrogen tanks, and breeding charts than if you were a passable heeler. Big and middling' money was getting back into ranching and the bottom line was everything. We then had salt blocks which were beginning to incorporate trace minerals. These brought with them on-going discussions about whether or not free-choice mineral access was better for cattle. Convenience in many cases won out and the salt block quickly became pervasive. We've all come to take them for granted, these hard heavy near-cubes with their locator holes indented in the bottom. Be honest now have you ever given a nearly-done salt block a second glance? Around Baker City, Oregon they do, and with several good reasons.

In that ranching and mining town, every fall for seven or eight years now, ranchers go out to pastures to gather what they think are their most 'attractive' nearly done salt blocks for the Great Salt Lick Contest and Auction. This is not a joke. This is for real and for good; people actually auctioning off the sad remains of old cattle salt licks. Results? Consequences? Baker City has found a humorous and righteous way to bring community together and honor its heritage while contributing to a most worthy cause.

In most parts of ranching country folks don't worry about someone stealing the remains of old salt licks. They can't imagine why anyone would want to. And the idea of forging a phony weak end to a salt lick is positively preposterous. In one darn sure western part of Oregon this is something which has changed in a big way and all because of one man. Whit did it. Whit Deschner of Sparta Butte Ranch. It was his epiphany, his vision, his sardonic sense of justice and elegance, his desire to turn a common oddity into an event that provided a feel good way for his community to come together and contribute to a worthy cause.

Baker City, Oregon, is cattle country. If you were to triangulate from Pendleton to John Day to Baker City, apologies to the great southeast of Oregon, you would find yourself in the heart of the inland empire of this rich and varied state. Mining, forestry, wildlife and ranching abound. The Snake River Canyon, the forges and easels of the great Wallowa communities of Joseph and Enterprise, the gold mines at Sumpter and Granite, the fisheries - elk - deer and yes the dangerous wolves, the deeply etched Native American traditions and history vibrating each year at the Roundup's Happy Canyon, the Chinese tragedy of the Lilly White Mine, the long loop traditions of Jordan Valley and the Owyhee tribal cowboys, Powder River, sagebrush, mountains and pine - it goes all in a swirl to make up this region, old and young all at once.

Out east of Baker City a bit is the remote community of Sparta. Whit Deschner lives there on his spred, Sparta Butte Ranch. Ask around Baker City about that 'salt lick guy' and you'll discover that this town is mighty fond of Mr. Deschner. They speak of him as brilliant, funny, caring, quirky all with tones that say this guy belongs to us, he is 'us'. Quite a testament but well-earned. Whit has Parkinsons. When we met last September he answered my question "What is this Great Salt Lick Auction business? How'd it start? What is it about?"

"I've always seen salt licks, hiking the rangelands, and I've wondered why this one, what makes this rare or special? I was sitting out at a cabin by the Lilly White Mine with a friend admiring the shape of a salt lick the deer had worked on and I began thinking that it sure beat some of the sculptures in parks and in front of buildings. You know the ones—the boulders with a chip knocked out of them masquerading as art that some artist has been paid a six-figure sum for. Not only were the animals creating these blocks not getting paid, but they were being eaten. In any case the idea of a contest formed and one thing led to another..."

Way back 6 years ago, Whit took this idea to a friend who owned the local coffee shop. Deschner said "What if we gathered up a bunch of donated tired salt licks and auctioned them off as art objects - to the highest bidder. Then we took that money and gave it to Oregon Health and Science University/Movement Disorder Clinic for their Parkinson's research?" His friend thought he was nuts. (Over time that same friend has become a prime sponsor and volunteer.)

Whit made the rounds to local businesses until he got folks hooked on his idea. Money was donated for prizes and ranchers were contacted and invited to submit their "best" salt licks. That was 7 years ago.

Whit roped in his neighbor, Nib Daley, a self-made auctioneer and local cowboy comedian, to call the sales events. At that premiere event 29 salt blocks were entered and the first one sold, under the auctioneer's gavel, for $129!

Over the years, with increased volunteer help and clever added aspects the event has evolved into Baker City's own iconic signature event, one with a growing worldwide notoriety. The posters say it all, 'The Great Salt Lick', Baker City, Oregon. Blocks have come from as far away as Las Vegas, and even included a salt block from Germany (submitted in photographic form).

This is the way it works now; Ranchers bring in salt blocks, with suitable entry forms filled out indicating what animals licked the block (i.e. cattle, horses, sheep, deer, elk, guinea pigs etc.) and a hand-writ poem commemorating the block. These are taken to the local feed store where the ranchers get a new one in trade (thanks to sponsorship support). Beautiful pine boxes, each just the right size for one salt block, are made by the prison and delivered mid-September to the auction volunteers. The local FFA is given advance notice to have one highly particular bovine heifer prepped for special duty sale night. The entered blocks are then "high-graded" to select the best ones for the oral auction (others get relegated to the silent auction). Whit designs interesting posters that get put up all over town. Sale night the winners are announced, poems read or sung, and the auction begins. Whit reserves the right to have the winning licks taken to the local Blue Mountain Fine Arts foundry, there bronzes are made for private sale the subsequent year. The winners get cash awards PLUS the bragging rights. All the auction proceeds go to the Parkinson's Research fund at OHSU in Portland, Oregon.

In 2011, the high selling block went for $525. Not including this year's event the total generated for OSHU has come to more than $30,000.

I asked Whit "What goes into the process of deciding what is a high-grade salt lick." His answer, "Its like pornography, you know it when you see it". Local ranchers in committee are the judges, one year the county agents did the duty. "It draws the whole community together. The ranchers come to town for this when they wouldn't ordinarily come in for art functions, even if it was for paintings of cows and horses."

I asked him, "Do the same ranchers come back each year with blocks, any new faces?"

Whit answered "Maybe ten to twenty percent turnover. Its become a competitive thing, they are competing against each other. We're beginning to see them steal each other's blocks. And there is a forgery division."

"You mean to tell me people are actually going to the trouble to create a phony salt lick?"

"Yes, but we can usually tell which ones they are. We put those in their own division. We have a best song division too. We also have an award for the one which most looks like Michael J. Fox. I did Janet Reno one year. She was really nice and asked for a picture. I think, (when she saw the picture of the salt block) I might have insulted her.

"Is it a point of pride for the ranchers to have the best block?"

"I think so."

"Are there those that swear that a Charolais does a better lick?"

"Last night we learned from an Angus breeder that the Angus do a better block. This year we are doing a song contest, there is an Italian family here whose son sings opera. He's going to sing the competitor's poems to old standard tunes like 'Sweet Betsy from Pike' and 'Yankee Doodle Dandy'".

"How do you deal with the case of a tie?"

"The FFA kids bring in their heifer to break any ties. We put a block on each side of the heifer and the one she turns to lick is the winner."

I had to ask, "Would you be able to tell if I had coated mine in molasses?" The look on Whit's face told me I had inadvertently given him a new idea.

"So, Whit, where would you like to see this going next?"

"I would like to see giant bronze salt blocks up and down Baker City's main street. I think it would be a wonderful draw - if it was right shape and large enough it might even double as a skate park. The library has always wanted a sculpture there, we're trying to raise the funds for that. I went before the city council and they are all for it but when it came to the high cost of the bronze casting ... they took it with a grain of salt."

For more information and a look at all the past winners visit www.whitdeschner.com

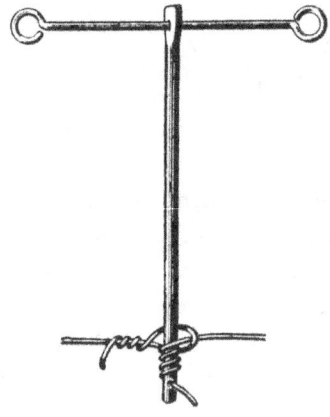

five

three eggs in a two egg pan

Around the world, today, hundreds of thousands of people have taken to the streets demanding that human society be put right. Media looks for a serviceable narrative in all of that, one which will give a longish run to demonstrations of discontent as reality theater, perfect for accelerated advertising sales. Stalled economy, wars, food poisonings, climate mayhem, poverty, hunger, disease, information anarchy, and a completely dysfunctional governance sector throughout the so-called developed world. Things are out-of-whack and getting worse. Several someones are pouring gas on the flames of this, our bad time. Money's being made on misery. Power's being gathered out of our demonstrated discontent. Rancid humor is being cultivated out of the decay of our society. Certainly a guarantee that people will keep returning to the plazas of our cities demanding that the cream-filled corporate cretins and the piss-tachioed morons of governance(?) step the heck aside. See in this the makings of a synchronistic global revolution which, if ignored, has the potential to flare into full and terrible anarchy feeding a plague of unanswerable sentiment and retribution.

But we err if that is all we see out there. These are also magnificent times. Look here, and over yonder, and behind that city edge, and atop that building - there are grand small life adventures, warm cool truths, musical fertilities all waiting to applaud and encourage our view towards a better humanity. We see glowing lights germinating into topside existence and growing all across the North American countryside. And most of these lights of endeavor have in common a bond direct and indirect back to the rebirth of good farming. And not just in N. America; all around the planet folks are taking charge of their own sufficiency with small farms and gardens springing up from Uzbekistan to Indonesia and on to Louisiana, from the south of Africa to Scandanavia. Delightful to imagine that we may be heading for a small farm earth. But we aren't there yet and many impediments are in the way especially if we turn our eyes and ears away from the roar in the streets.

Good farming, that hand-held sensorial salad-like agriculture resident in our genetic memory, is today's unsung best hope in the swirl of societal options. It offers up right livelihood (jobs we might embrace as truly fulfilling and sustaining) to hundreds of millions of people, it provides a catacomb-like community of little laboratories where the fertility and diversity of biological life bubbles over into a shared over-lapping regenerating abundance, an affordable wealth. It is what most of us want for ourselves, for our times, for each other. It is what only a few of us believe is attainable. And that is not because it is difficult or out of reach, it is because we are told, indirectly and repeatedly, that it can't be done, and we haven't been able to hold off that negative message.

One of today's greatest threats to good farming is our run amok journalistic community. Bandied about for decades has been the bizarre notion that journalism is a profession leaning heavily on the best of ethics; that journalism somehow owns the moral high ground; that we may trust most of the journalistic community to be looking out for our collective best interests. It is most certainly not true, not now and not ever. Journalism depends on the formal pretense, to appear to be objectively reporting the news while inciting and exciting audience and readership with petty peculiarities, vulgar extremes and terrible announcements. Add to that the obvious handmaiden-relationship media news has to corporate governance and it is easy to see that for them telling a clear-eyed story about small farms and good farming is either silly or dangerous or both. Bad enough on its own this murky sludge-frother we call mainstream news, but today we have thrown into the mix hundreds of thousands of instant expert snobs-without-portfolio availing themselves of social networking and internet 'blogging'. We are in the midst of the 'thought' equivalent of a land rush. People are flinging themselves, pell mell across the cyber-scape driving proverbial 'stakes' into the internet, claiming that this or that thought cluster is theirs and theirs alone.

Why does this matter to us? It matters because this so-called "free flow" of information (read 'my-wednesday-thought-on-the-subject') is like a river of poisonous wet concrete, it surrounds everything in our society and threatens to kill off our possibility and hope as the mixture dries and hardens. It matters because the din of public mutterings about what is inevitable, what is hopeful, what is possible is thinned and even occasionally snuffed out by these self-ordained social critics who insist that small independent farming ventures, that the life and work we cherish and work to put forward, is archaic, irrelevant, romantic, nostalgic and doomed.

My gosh but we have a lot of people out there with way too much time on their hands! What a wonderful world it would be if they had to provide true community service before they were granted a license to mutter publically? What flavor would this endless river of "blogging" take on if each and every one had experienced, first-hand, life and death on the farm, apple harvest, manured splinters up finger nails, the sweaty invigorating fatigue of the harvest just in before a storm, the sight of every single young carrot lying withered on the surface of the garden with your own child so proud of her weeding job?

Small Farms mean business, they mean food nearby, they mean hope, they mean opportunity, they mean jobs, they mean healthy small towns, they mean a healthier planet, they mean a way straight out of the darkness. Why is that story going unreported?

I was recently listening to a classical radio station program featuring keyboard music and the commentator offered "If more people learned how to play the piano there would be less war in the world." I had only to pause a few moments to realize the same could be said of farming and gardening. A friend remarked recently that she heard a Chinese news show via the internet where the commentator, interviewing an editor with the Wall Street Journal, asked "Why did you wait three weeks before reporting on the 'Occupy Wall Street' movement?" The editor had no answer. In other countries, whether we like it or not, folks read of how it is that millions of U.S. citizens are out of work, many homeless and hungry while our government, at every level, works feverishly to protect the property rights and income of the rich and powerful. Our friends in other countries are not surprised that most major US cities have "occupy" demonstrations demanding change. How is that any different from what is happening in the middle east? We commented before on how the media pushed the story of the Arab spring revolutions as having germinated from social networking on the internet and giving barely a nod to legitimate issues with hunger and poverty.

Affordable wealth
Aligned fertility vs alleged efficiency?   What price balance?

We must tell our own stories to one another. And make certain they are heard.

A few years back I visited Spence Farm in Illinois and spoke with Marty, Kris and Will Travis - the eighth-generation to work and live on that same acreage. At one point Marty shared with me a snippet as a piece of the wider puzzle of Spence Farm's successful, phenomenal and important diversity. They harvest wild (and encouraged) Nettles to sell to restaurant chefs who use them as a cooked green, like spinach. On the side as it were, that's $480 per week added to the usual farm income.

Just before Halloween my friend Mike took his Belgian team and people hauler to the neighboring pumpkin patch to give rides for the weekend. The pumpkin farmer advertised rides and throngs of folks came out to do just that and get their holiday sustenance. Mike said that well over a thousand people paid $2 a head to get a wagon ride, all gravy above and beyond the sales of the big orange globes. Once again, on the side as it were...

These are examples of how and why it is that we feel so optimistic about the future. Examples because they ride above and to the sides of the real work offering a teaser of those myriad ways, through creativity, that we can add 'aspect' and income to our good farming adventures.

So, you may observe "those are exceptional circumstances, you can't use those to make an argument that farming as a culture is viable. You need to restrict the discussion to the empirical evidence of efficiency and profitability as pertains to actual farming practises, not to silly side-dishes and goof-off stuff like wagon rides!"

Every single piece of the puzzle is a piece of the puzzle. Throw one piece out because you don't want to accept it and the puzzle will always be incomplete. Why do we not want to see and embrace that? We live in a time and space that grants to anybody the right to cast suspicion on example. Okay, let that be, but I refuse to allow anyone in my presence to deny, uncontested, the applicability of an "example" just because it would appear to some to be exceptional. Every single piece of the puzzle is a piece of the puzzle.

Ah, but you throw up your arms and say "it's about the evidence! Why can't you accept the clear evidence that ..."

Forgive me a true story by way of answer: I have a good friend who is a hay farmer. As a young man he lost one eye in an accident and he wears a glass eye in its place. One day a few years back, he and another man were working on an enormous haystack when it collapsed on my friend and rendered him unconscious. When he came to in an ambulance the medics had strapped him to a back-board and were checking his vital signs as his worried daughter looked on. He slipped back into unconsciousness so the paramedics took a light and peered into the wounded fellow's eye to measure its response by dilation. The evidence was clear and required immediate action! When the paramedic saw NO response in the eye to the light he yelled for the shock paddles and both medics prepared to try to electrify this fellow back from the dead only to hear the daughter screaming "NO, NO, check his other eye!" They were mistakenly checking the glass eye and thought, because there was no response, that his heart had stopped. Had the daughter not screamed there is no telling what damage might have been done in the electrical effort to jump start a heart which didn't need a jolt, it was beating just fine.

That story works to make my point; what appears to be clear evidence might be something else altogether. But that tale also works on another level, as an analogy for how we see small farms and the family farm in general. For decades we had been told that the family farm, the general farm, the small farm were all dead, relics of a distant past. We were told by industry, academia, and governments that, though they may be deserving of protection as cultural relic, they certainly didn't offer any realistic contribution to the larger challenge of feeding the world. (Oh, how miserably wrong they were.) In those quarters today food production is celebrated as a vast and vertically-integrated industry critically dependent on ever larger doses of chemicals, high-tech implements and the magic(?) of bioengineering. But they've missed the party, and they've missed the memos. Most important they messed up the diagnosis. The patient didn't die. Small Independent farms ignored? Yes. Abused? Yes. Trod upon? Most certainly. But not dead. And this same community of small independent farmers has quietly sprung up full-strength giving birth and purpose to a whole new wave of good farming. Tens of thousands of new small farm ventures are germinating all over the US. New small farm ventures embracing new and old practises and making profits while reconnecting local populations with the culture of agriculture. Good farming is the fastest growing sector of agriculture world wide. Industrial agriculture is dying by its own hand as "pesticide-ready" bio-engineered corn falls over in the field, as hormone-laced milk is turned away at the marketplace, as massive corporate ag companies struggle to compete while their wonder crops and chemicals fall prey to greater pests and pestilence.

The timing's right. These are bad times. And good farming offers the antidote. With good farming the attainable and worthiest goal is always for affordable net gain: net gain in fertility, in the strength of biological diversity, in caloric production, in overall plant and animal health, in environmental improvements, in water quality and retention, in yields, in jobs, in small towns, in neighborhood, in community.

Why do I keep sticking into the argument that word "affordable"? I do it because I believe it is the critical missing adjective, it's left out of most people's wish list. "How do you have a net gain that is affordable? How do you measure something like that? Why should it matter?" As we work to gather unto ourselves the 'rewards' we pursue, do we ask repeatedly whether or not these 'rewards' cost us too much; cost all of us too much? We should ask these things, it should be a piece of our morning prayers. The small farm would teach us that there is the inherent potential for us to feed the world well without mining the planet's fertility and killing off her biological diversity. If we take care of our farms, they will take care of us.

Two decades ago, in Nevada, I had a lively discussion with Allan Savory the sage instigator of holistic resource management and impact grazing. He made a point at that lunch that has stuck with me all these years. "Your objective should be to 'raise' better soils, if your fertility is improving that is your true net gain. To that end hay and grains produced should stay on the place to feed back into livestock and into the soil itself. Livestock should be seen as an instrument towards building better soils. If you find at the end of the cycle you have increased your cattle numbers and have some to sell, it is almost as if you are finding yourself with a waste product that folks will pay you money for. You should see the money you receive from the sale of those cattle as a reward for the real net gain you accomplished - your ever improving soils."

Affordable wealth? Is it possible? You betcha! Wow, what a world that would make!

And it's all available, all possible, all necessary. It starts with millions of small independent hands-on farms. And the great news is that millions are already in place ready to be joined by millions more. They just need to be told that what they do is valued, important, even critical to the life of the planet and humanity. They need to be permitted to do the beneficial work.

I enjoy cooking on cast iron. We have a small skillet, just perfect for frying two eggs. When eggs are abundant and I am in an early morning hurry, I will sometimes use the moment to reward myself with what I see as an extra unit of our affordable wealth. We get our eggs from Brian MacNaughton and A.J. Ferris, on their separate farmsteads the chickens are, dare I say, coddled. The resulting eggs are magnificent, many a time double-yokers. And they are various; various flavors, colors, sizes, thickness of shell. I break open two eggs into the waiting butter-lined hot cast iron pan and then smile to myself as I take a third and add it to the crowd watching as the two are squooshed over to make room for the abundance. Yes, there are days when one egg does me just fine, others when I do without. But see with me for a moment the picture of a unit of measurement, in this case a certain size of frying pan perfect to accomodate two eggs - so we call it a two egg pan.

'Agriculture' (the industrial model) I see as represented by two eggs in a two egg pan; that's the maximum equation. No questions. Clearly (to some) that is what fits. Clearly that is the target. When you get two eggs in a two egg pan you may consider yourself an efficient and successful industrial agriculturalist. How could there be more? You couldn't put three cups of liquid in a two cup measuring container, could you?

Many will nod at my physical form and argue that I am the last person to need three eggs, and I will bow my head and sometimes aknowledge this. But that is not my point. The point I wish to make is that "good farming" regularly gives us opportunity for the sort of abundance that is, if we can see it that way, merely an indicator that all is right with our world, right and getting better with each passing moment. Those chickens are healthy and happy. The eggs we take from them give us sustenance and more reason to make certain of the poultry heaven.

If you are growing pumpkins, the oft-referenced USDA enterprise data sheets will tell you that all you can produce on your soil and with your climate is X number of pumpkins which might reasonably bring X number of dollars. Finite? Fixed? Clear? BUT what about adding aspect to the enterprise as in the case of Mike coming over with a horsedrawn people hauler and giving wagon rides to folks who come to purchase pumpkins? You may find yourself, as an old Missouri pumpkin-farming friend once did, having to purchase pumpkins from the neighbor's farm to 'reseed' your own in the evening because that first day all you actually grew were sold. (That man also used as an advertised attractant horsedrawn wagon rides out to the field for U-pic conveyance.) I say adding aspect is a clear sign of good farming, of three eggs in a two egg pan. I say the secret of success lies squarely on the systemic formulas each of us come up with to add aspect to our farming at every level.

And let's say you are producing maple syrup, meats, heirloom corn and other crops and you allow that your woodlot and field-margins harbor nettle plants which you carefully harvest to sell to restaurant chefs. That also is a case of three eggs in a two egg pan. Sometimes the aspects we might embrace and add have always been there waiting for our discovery.

In both cases the added aspect takes NOTHING away from the farm's fertility and adds to the diversity and elegance of their examples not to mention the bottom line.

Good and "care-full" farming will hold us true, and that becomes more real with every passing day as we humans are forced to watch and sometimes join in the terrible slide of our economies and societies into "primetime" anarachy. But we needn't accept that anarchy. We do have a choice.

Forgive me this prayer: Give us this day our daily hope that in all things living we see some goodness and we see ourselves. And grant that we might find the lights and use them in a walk to affordable health and wealth for everyone.

six

time to farm

It is time to farm, right now. Don't put it off any longer. It's time to farm because the opportunities for success abound. It is time to farm because many of the critical resources for the beginning farmer, pure parent-seed and livestock genetics for example, will become harder to acquire. It is time to farm because new farmers are the only answer which will ward off accelerating development of fragile farm land into housing. It is time to farm because communities are searching for locally produced food.

Do you know where your food comes from? Probably not, at least not all of it. Does it matter to you? For most people the answer is no, it doesn't matter. But to a rapidly growing number of folks the question of where their food comes from is crucial. Should it be important? Yes, absolutely, it is terribly important. Our health, well-being and connection to life itself are all reinforced by the easily accessible view of who produces our food, our knowledge of how it is grown, and our belief in the farmers who are responsible. Comfort and assurance an extravagant notion? I think not. Quite the opposite; it is extravagant to close your eyes to where your food comes from.

Would you prefer to purchase food from good conscientious farmers like a Lise Hubbe or an Anne Nordell or a John Davis rather than a large indifferent corporation like a Monsanto or a Cargill or a Phillip Morris? If you ask to see how the vegetables are grown at the Blue Hill, Maine Farm of Elliot Coleman you will likely be engaged in an experience that will add assurance, flavor and purpose to the produce you purchase from him. If you ask to see how a large corporate factory farm produces your food you will likely hit a stone wall where practicality and confidentialty will make it impossible for that experience to occur.

Imagine this; I hand you a sealed plastic bag with lettuce inside and a tag on the outside featuring a USDA endicia, plus contents listings and liability waivers. The address printed on the bag is a corporate office two thousand miles distant from you. Now, in an act of blind faith, you will take that lettuce from that bag and feed it to your child. You have no idea what chemical residues are in the bag, farm chemicals which are regularly poisoning field workers. You are asked to believe that your government and its department of agriculture are protecting YOU and YOUR child and guaranteeing the purity of this product.

Now imagine this: Lise Hubbe smiles and hands you a head of lettuce and cheerfully supplies anecdotal information about its variety, her farming practises, her work horses. She tells you stories about her farm. You are given access and connection. All of this gives you the accurate and comforting sense that your children and their children are protected and connected.

Our choice might be simple:
Pay the corporation for the questionable food product and take the risk that you poison your family and financially contribute to the continuing destruction of the planet and society
or
Pay the independent local farmer for the assuredly tasty and healthy food product and contribute DIRECTLY to the health of your family, the success of that farm, the improvement of your environment and the reduction of the coffers of the big corporation.

The choice might be simple, but all of this presupposes a vast and growing array of independent small farms. That is part of the reason why I argue it is Time to Farm, Right Now.

In April of 2006 we Millers traveled to Millersburg, Ohio, to share in the wedding of our dear friends Joe and Ruth Raber. For twenty-six years we have been making regular trips into the Wayne, Holmes and Tuscarora county Amish communities of Ohio for pleasure, for friends, and for business. We've wandered into many corners of that countryside and long been thrilled and delighted by the evidence of what a closely-knit community of free-holding farmers can accomplish. Beautiful farms, lovely crops and strong healthy livestock. Farms trading product and service, one to the other, in regular weekly markets exhibiting deep vitality.

But to the frequent visitor it is impossible not to record change, evolution if you will, to greater diversity of venture, and to a larger dependence on elements outside their community. There are increases in non-farm, home-based craft ventures and cottage industries now with what seems like fewer traditional farms. Social and economic pressures have more and more Amish working for industry. At least that is the way it seems. Perhaps land development pressures along with the recent decades long history of depressed crop and milk prices took their toll. (At least one sociologist claims that there are no fewer Amish farms. He argues that it appears that way because the community has grown in numbers resulting a shift in percentages.)

My April 2006 discussion with Wayne Wengerd of Pioneer Equipment began about their new plow design and came round to this subject as he expressed his concern that large numbers of Ohio Amish farmers are quitting the time-honored vocation to move to light industry and craft. This, he continues, at a time when we desperately need MORE farmers, not less and a time when opportunities for genuine success with locally-based farming have never been better. I agree with him on all fronts. (As an aside: It is an interesting paradox that even with these concerns and observations, Pioneer Equipment, a horsedrawn implement company, continues to increase sales every year.)

For thousands of years a large segment of humanity has felt naturally drawn to the farming vocation and life. It is in the blood, it is a piece of our genetic code. We feel a comfort, vitality and sense of purpose working with the land, with soils, with seeds and plants, with livestock. It has always been a bonus when we find along the way that we have produced something of quality with those efforts. Imagine good work which results in food and fiber and then as a bonus benefits ourselves and others?

WE NEED MORE FARMERS AND WE NEED THEM NOW
With all the illusions of a life of ease and convenience, why would any young person consider getting into farming today, at the front end of the twenty-first century? And would we, should we, as adults and parents who care about the best future for our young people, would we, should we, wish such a life of hard work on them?

And how about those adults who want, with or without farming in their history, to get a place of their own to raise livestock and crops?

As this society, in this time, when we pay close attention to the world and those around us, we have a pretty clear understanding of what our society needs. We need thousands more doctors and nurses who want to help people and prevent illness. We need thousands more dedicated, intelligent, curious teachers who want to teach. We need people in positions of power and influence who care about the planet and our environment. We need dedicated, intelligent, reluctant leaders, who, in spite of how busy and successful they are in their chosen walks of life, are prepared to give of themselves to work selflessly and 'part-time' in governance. You can continue the list. But make sure that you add this; we need <u>millions</u> of intelligent, dedicated, craft and soil-based, independent farmers who want to raise good food and strong families. Yes, millions. The planet needs millions of true farmers to feed us all and to keep the waters clear, the soil fertile, the biodiversity vibrant, the lights on, and our governments and industry at bay.

For farmers in America, it is the worst of times and it is the best of times. Industry wants us out and those who need food to eat want us in. Government no longer sees a value to what we do and yet the environmental future of the planet depends in large part on us. The costs of industrial farming are skyrocketing while the prices for industrial ag products remain flat. The bureaucratic roadblocks to an independent craft-based natural family farming operation are piling up faster than beer cans at a Nascar race, while the prices a concerned public is willing to pay for genuine food, produced in a healthy manner, are substantial and increasing. The growing (actual and/or manipulated) scarcity of petroleum product will soon put a crimp on intercontinental as well as interstate food shipments, which is bad news for food prices, but it translates to an enormous opportunity for small independent farmers who are intelligently focusing on local markets.

It is the best of times, it is the worst of times, to borrow from Charles Dickens. And the analogy goes deeper yet for there are many of us who believe that efforts are underway to guillotine true agriculture.

Who's the enemy? Do we have an enemy? Are we the enemy? Who or what is responsible for the agribusiness mess we seem to be fighting? It's too easy to point to the multi-national corporations who are working voraciously, if clumsily, to try to control all aspects of food production on the planet. Their motives come down to a grab for the power to control markets and thereby profit. We know where they are coming from and we fear where they are taking us. But what they are doing would not be possible without the aid and comfort of a well-oiled, 'professional', store-bought, government. Yet the real meanies, the real villains in this affair, aren't those elected morons of shellacked insouciance, of girdled intellect, as well as vulgar untrackable ancillary income. The real bad guys are the architects of our agribusiness system, generations of stiff-necked bureaucrats, statisticians, ag economists, and program directors with the USDA, career professionals who have been high-jacked by industry to do its bidding right or wrong. From their closeted, disconnected, mean-spirited myopia they have worked relentlessly to reduce agriculture to its lowest common denominator - production. They have been working to get people off the land and render farming into a purely chemical/industrial equation.

I have argued that we need to abolish the USDA. I still feel that way, but perhaps we don't need to take any action. Perhaps they are doing it for us. They are dinosaurs which have blindly set in motion their own obsolescense. A behemoth bureacracy like the USDA requires a constituency. They need a base of people to serve. As they drive true farmers off, they lose their base and very purpose for being. As they lose credibility with the consuming public they also lose base. This, and the fact that young college graduates are less and less inclined to follow the bureaucratic line and we can see the handwriting on the wall for this relic of the past.

If a young person is considering a career in agriculture the WORST thing they could move towards is working for the USDA. There is simply no future, honor, or reward in such a path. If a young person is considering a life in agriculture the BEST way forward is with an independent farming operation of creative design and economics. Such a path offers hard work along with honor, reward and the best of futures.

WE NEED INTELLIGENT, CAUTIOUS, DOUBTING FARMERS, RIGHT NOW

If pigs, sheep, cattle, horses, goats, chickens, etc. figure into your dreams of a future of independent farming, DO IT NOW! DO NOT WAIT ANY LONGER. If you are considering expanding your farming operation with expensive and/or difficult infrastructure such as processing facilities, new livestock housing, refrigeration, and delivery systems, extra thought needs to go into how the proposed mandatory, federal, animal identification program might hurt your investments. It is a time to keep our eyes wide open and think, clearly and deliberately about what it might mean to have state and federal government, in concert with industrial subcontractors, knowing who you are, where your buildings and facilities are and where all your animals are at every moment. I find such a concept to be not just meddlesome and scary, I find it to be sacrilegious.

If the National Animal Identification System becomes law anytime soon, and unfortunately it does seem possible (it's not a done deal yet), it will make it difficult or impossible for anyone to acquire livestock without entering into that system. And complicity with that program will register you and your farm with the government, which can and will make regular defacto determinations about your right to continue as a farmer. Many small and midsized livestock support companies have indicated that NAIS will, if it is implemented as mandatory, effectively shut them down. That includes hatcheries, livestock auction barns, rare breed farmers, small dairies, custom meat processing plants, etc. The mixed livestock operator who starts today may enjoy the last opportunities of truly free access to a diversity of breeds and individuals not registerd with the government.

In our work with *Small Farmer's Journal*, we have gone on record as being completely opposed to the NAIS. We know, through channels, that tons of letters and phone calls were placed with state and federal officials voicing specific opposition. We are proud of those who made the effort. It is easy to be a bit stymied by the appearance that these letters have elicited only frustrating doublespeak response from the USDA (i.e. "don't fret, for the time being it's only voluntary, as long as we get enough compliance ...") and not much from the electorate and appropriate agencies. In spite of these appearances of no gain, voiced opposition has actually had a tremendous influence on the future of NAIS. The vulgar propaganda dogs of agribusiness may be in retreat. Government and industry are scurrying to put out the fires we are setting, struggling to regain what they thought was their moral certitude on this, and similar issues.

Remember:

1. A national animal identification system will NOT safeguard the food supply infrastructure in the US from contagious disease.

2. The oft-named diseases of concern - avian flu, mad cow disease to name but two - can be monitored and controlled by existing regulations IF the USDA would force factory farms and processors into compliance.

3. Using the "terrorism" card to frighten people into thinking they have to comply with yet another constitution-busting surveillance program smacks loudly of police-state tactics and fascism.

4. The costs and procedural nonsense required of such a program would effectively shut down what is arguably the most important segment of our agriculture, the small independent farmer.

5. All the various aspects of this proposed program, if implemented, would give additional competitive advantage to the largest livestock operations.

We know, and have shared with Journal readers, that several organizations are working very hard on this front. Their work is vital. It is important not to be discouraged by this animal ID end run. Vigilance. We are not done yet. Continue the fight and encourge others to do the same.

We need MORE intelligent, cautious, doubting farmers, right now.

SOCIAL WARMING

I am one of those who believe that the phenomenon we call global warming is a very real consequence of humanity's industrial footprint on the globe. I also believe that the vast majority of the human race doesn't give a fig about global warming. They could care less because the consequences, if there are any, seem to be far removed from their connected lives. But what I believe doesn't matter. It's what you and they believe that matters.

We humans seem doomed to repeat ourselves. And the evidence seems clear that we are entering a new dark age with distances between peoples and cultures increasing rather than decreasing. In spite of the insistence of the stock market and industrial propaganda pointing in the other direction, we are watching the beginning of the end of so-called global free trade. The future, good and bad, will belong to regions and there the character of local self-reliance will rule the day.

Our old world recently came to an end and we missed the passing. The end came with the internet and now wholesale criminal, governmental and industrial access to individual computer data records, with the USDA usurpation of organics and the slithering forced insertion of animal ID, with the declaration of war on a noun (terrorism), with the political development of the rule of deniability (steal, cheat, lie, and then just say it isn't so), with the absurd trivialiization of the US Supreme Court, with the media revelation that most organized religions have been protectorates for many forms of deviant behavior, with the discovery that it takes very little to purchase a vote in most any government, and the list goes on. It would seem that the only thing left for us to discover together is that "political correctness" is the ultimate corrosive and useless oxymoron. Even with all of these indications, society seems to rush headlong BACK into the abyss, back to the internet to divulge more personal information, back to allowing leadership to consider "regime change" in other countries, back to sanctifying the lying thieving congressmen and senators, back into the dirty bed chambers of organized religion, back into store-bought government, back into a head-in-the-sand posture. All of us, perfect saps for the market place.

Because of it, we are at risk that industrial propaganda and fascist reasoning will win the day giving us individual, plastic-wrapped, irradiated salads with synthetic cheese and attitude-adjusting farm-aceutical dressings all for $23 a serving. And these will eventually be sent electronically, as chemical equations, direct to little transporters located in our kitchen-less homes.

The antidote is to have more good farmers producing real food for their neighbors. Those same farmers will begin to heal the environment in small ways right on their own places. And those healing actions will join together in a fabric of fertility and self-reliance beginning a process to soften the footprint of humanity on the globe.

It's a hot time to be a human on the planet earth, and its getting hotter every day. Those of us in so-called developed countries are hot (read angry) because our neighbors ain't behaving the way we want them to. Shame on us, bloody shame. There is an intensified focus on the little differences between individuals and groups and zero tolerance for diversity. We don't see a child, we see a Muslim child or a black child or a wet-back child. We don't see a mother we see a welfare mother, we see an unwed mother, we see a half-breed mother. We don't see a free expression of ideas, we see the targeted agendas of demons out to destroy what we believe in. (The joke in that is that many of us don't actually believe in anything, EXCEPT our absolute right to have it our way.) The agendas of the Granola, Grits, and

Martini sets have decided that no one is going to take their health club, church club or golf club memberships away from them, NO ONE. Environmental actuarials don't matter, pandemic disease doesn't matter, food security doesn't matter, they don't even seem to care what they might have to pay for gasoline - all that matters is their absolute right to their particular version of socialized pleasure. The distinctions between freedom and entitlement have vanished in the western world. Meanwhile, quite to our dismay, the rest of the world - and last time anyone looked there are far more of them than there are of us - the rest of the world is heating up for opposite reasons.

Every three and a half seconds somewhere on this planet a child dies of hunger. THAT should matter. The climate of the planet is altering at an alarming rate and life is being threatened. THAT should matter. Devastating wars are being fought to satisfy the twisted, ego-maniacal, shortcomings of little idiot cowards in power (from power-crazed female New York senators to ego-maniacal matinee governors to small-brained appointed presidents). THAT should matter. Industry is shutting down farming. THAT should matter. Instead we choose to snipe at one another about real and imagined sideways glances construed to be acts of disrespect and threats to our rights to excess. We yell at each other from one SUV to the other. Here at home in the West we don't like each other very much. Suspicion rules the day.

In the so-called undeveloped world there is an intensified focus on the wholesale difference between the haves and the have nots. There they see hungry children and imagine those OTHER children riding in bodyguard protected tinted-glass limousines. There they see hungry struggling courageous mothers and they imagine the kept, sterile, brainless women belonging to the leaders. There they see hardscrabble farmers scratching the dirt to get a tiny piece of food and they imagine huge warehouses full of magically appearing processed foods. Out there, injustice is creating pockets in shrouds where weapons are hidden until the right moment. Its a hot time, and that is not good.

The antidote again, even out there, is to have more good farmers producing real food for their neighbors. Those same farmers will begin to heal the social and physical environment in small ways right on their own places. And those healing actions will join together in a fabric of compassion, justice, fertility and self-reliance beginning a process to soften the footprint of humanity all over the globe.

Faith has become such a maligned commodity in these times. We are asked to have faith in our federal and state governments, faith in our industrial food supply, faith in the good intentions and objectivity of mass media, faith in our laws and court system, faith in a concept of man's basic goodness, faith in our health care system, and so much more. Maybe it's screwy but I always assigned to my definition of faith a required measure of belief as in "I have this faith because I believe..." I believe in this country but I no longer believe in this government. I do not believe in the industrial food supply. I do not believe in the mass media. We need the rule of law, but I can not believe in a justice system which rationalizes ideologies, panders to the powerful, and regularly squashes the little guy in order to protect the property rights of thieves. I do not believe in our health care system because it does not care for health it cares for profitable diseases. I want to believe in the basic goodness of man but these times sorely test even that notion.

Where then do I justify "faith" in any of these things? I say circumstances point to constructive doubt and even suspicion, I doubt and suspect our governments, our industrial food, our media, our justice system, and so forth. Until I can believe in these things I will grant them none of my faith.

I do believe completely in the concept of the independent small farmer. I know what he and she can do. I am regularly thrilled by what many of them have done and can do in concert. <u>Because</u> I wholeheartedly believe in true farmers, I have faith in them and what they represent. They represent for me a real opportunity for a better world.

How do any of us make a difference? How do we start to turn this stranded, wallowing ship around? We do it as individuals working in constructive ways on those things within our grasp. We do it with small steps. We do it by encouraging and assisting others to join our ranks as farmers. I believe we do it by being the best farmers we know how to be. We take great care with each and everyone of those daily choices we must make; choices of what we are doing, choices of how we will do what we are doing, and all of that supported by a gratitude and appreciation for why we do it.

It's time to farm.

seven

drumming

The answer is; "men, women and children on their own piece of dirt, building sweet fertility while raising animals and crops to feed people."

The question is; "what on earth can we do to save the planet and humanity from extinction?"

There are many who can't accept such a simplistic notion. Some don't believe that people and the earth are in jeopardy. Many who do believe we are at risk cannot see beyond their chosen cause or focus; be it greenhouse gases, environmental pollution, animal rights, food safety, farmland preservation, endangered species, hunger, war, pestilence, disease, corporate malfeasance, or moral degradation. In each and every case the fight has often been to treat the symptoms not the cause, (an approach which has given us a failed health care system). This seriously aggravates the problems as it has been applied to the gamut of these other concerns. For example, biodeisel will not save us from global warming, it will hasten it as it speeds the poisoning and depletion of our soil resources. A centralized data base maintaining the whereabouts of every small farmer's poultry and goats (with colored astericks for location of feedlots and pig factories) will not protect the food supply from pandemic disease, bacterial infestation, or terrorist infiltration. It, instead, will increase the likelihood of these problems by providing dangerous exemptions for industrial producers and creating a readily available computerized roadmap of our food system frailties. Setting farmlands aside in park-like mode does not preserve them as farmlands. Instead it adds to the net loss. And yelling and screaming that the fighting must stop while refusing to see that <u>our</u> economy thrives on the war profits is like facing a household fan into a high wind.

"Enough! Nobody wants to hear such things. It is instead, time for the comic relief. There is just so much of this doomsday rhetoric that we can stand. If you have something constructive to say, say it in a pleasing way with pretty pictures; make it diversionary, give us a reason, if only for a moment, to believe you."

SOMETHING CONSTRUCTIVE TO SAY

Many of the best small farm examples I have discovered through my work and travels I am unable to share or showcase because the people insist that their failures, successes and secrets not be publicized. I choose to honor that. Instead, here, I take liberties melding the disguised elements of 3 very real farm families into one fictitious narrative which I hope has some demonstrative merit.

On Dan's farm, just thirty acres, the decision was made to change out the greenhouse heaters from propane to wood fire. He worried just a little that maybe the wood smoke might be a pollutant until he remembered from years of experience, that the design of the wood stove as well as the nature of the cordwood, all had dramatic effect on the smoke outlay. He had the excellent firewood held in a common woodlot with his cousins. The propane had to be bought and delivered.

Besides, the new plantation system in the greenhouses, with heated water ways and layered composting manures collecting and creating warmth under the secondary tenting of plastic, was requiring far less auxillary heat. Ever since he had read about the Coleman's portable greenhouses and employed some of the plan to his own flower and vegetable production, it had become second nature for him to see how new ideas might relate. He had come up with his own homemade facility and system for dragging the greenhouses along an extended bedway and incorporating a netted framework for the poultry. Two greenhouse structures and one poultry structure, each thirty foot by sixty foot, were placed all in line on three hundred feet of bedway so that every other season the poultry were contained on a former plantation bed to eat all the insects and larvae and fertilize, worked slick. It took three years, some mistakes, and a chunk of change but it was all worth it. He marveled at how the soil kept improving and wondered if there was any end to that incline.

Whenever and wherever, Dan did his own cobbled-together constructions. For him, the bigger part of success was in controlling expense. One of his proudest achievements had been the portable waterheater for the greenhouses. On a heavy-duty, steel-wheeled running gear, he had built a metal platform with hinge halves all around the edges. He set an old waterheater tank in the center of the platform. The tank had pipes running in and out but no obvious power or heat source. Then he affixed old oiled, cement-form plywood to the hinges to form a box around the base and tank. He parked the apparatus near the greenhouse and filled the box around the tank with fresh manure from his stabled workhorses. As the manure het up, it warmed the tank and water which Dan used in cold weather to fill heat transfer ditches in the greenhouse. When the manure was composted and no longer generating heat, he dropped the plywood sides and pulled out the compost for the greenhouse, moving the machine to its next location, folding the box back up, and loading with fresh manure.

Dan's wife, Mary, was happy to let her husband be the mechanic inventor of the farm family. Better this than the family history of distilling corn liquor. Herself, she preferred the marketing, all the way from harvesting and packaging produce, eggs, meats, and flowers to the three different farmer's markets she attended each week throughout the season. She worked as a credentialed substitute school teacher. Even so, in the winter time she always found herself itching to get back to the market stands and her hundreds of devoted customers and friends. There was a conviviality and exchange that filled her up so much that she secretly knew she would have to do this even if they didn't make money at it. But they did. They made a steady and ever growing stream of money. Plus, recently it had been a wonderful enterprise to share with her growing daughter.

With Chip grown and gone on to his teaching career, Dan and Mary had wondered if they could keep up with all the work but so far, Calley, their daughter, had picked up the slack in most ingenious ways. Like when she had convinced her dad to let Bingo, her trail horse, learn to pull a small one horse spreader through the greenhouses to spread compost and amendments. It was far handier than the Brabant team of draft horses. Plus it had made little Bingo into a much better horse all round. Once the gelding had learned to accept the squealing aromatic piglets on the one end of chores, all the way through to calmly walking into the sometimes-slapping plastic hoop house, any other threatening experience seemed like child's play to him. And Calley gained confidence along with her horse. Yep, Calley had become a real mainstay. Only thing she couldn't get into just yet was the chicken butchering. Loved the chickens, even named many of them. Hated the butchering, but she understood its place in the scheme of things.

Dan figured they had about an acre either in greenhouses or affected by them. And then there was another four acres of intensive open air market garden. Tomatoes had been their top money maker for a long while, but specialty herbs and cut flowers were, to his surprise, gaining fast. His twenty acres of remaining arable land had been divided into five four-acre fields which he religiously rotated with occasional crop variants. From pumpkins to watermelons to buckwheat and field peas, he loved the challenges and the variety. He wasn't sure if he was doing the right thing by pasture and hay crops as he harvested something from every field each year. His son, the ag teacher was telling him he should think about keeping some of his land in a true permanent pasture. The way things were he usually had three fields in legume/grass mix, alternatively, for a three to four year run until he plowed it up and planted grain and then a specialty legume, squash, or melon patch before returning to the legume/grass hay and pasture mix. This had worked extremely well for them. Employing the lanes and four acre woodlot on occasion as horse, pig and cattle pasture, he had been able to take a hay cutting off the fields before using them, subdivided by electric fence, for rotated stock pasture. And there had usually been some feed value to the other rotated crops as well. He usually put up about twenty five tons of hay each year which took very good care of the winter feed for three horses, plus one milk cow and six beef cows and their calves. Four acres of grain production went a long ways towards helping keep the poultry flock which, with the pigs, enjoyed many advantages from the garden surplus. Their dozen bee hives were a bonus all the way around. Nothing went for loss.

Dan had enjoyed converting a small old Allis Chalmers motorized combine grain harvester into a fairly modern stationery thresher. He had cut the conveyor bed off an old ensilage blower and attached it on an incline to the throat of the combine allowing him to throw bindered bundles into the machine for threshing. He did this so that he could continue to run his favorite implement, the old McCormick five foot grain binder. His other implements, mostly horsedrawn, ran the range from new to old, each and every one debt free.

They weren't eligible for any kind of production credit lending which was fine by them because they didn't need it. Mortgage on the land was all paid up. Property taxes were all they had to pay that way. In fact the only monthly contract payment they had was on the van Mary used to go to market.

In the beginning, twenty years back, they had gone against conventional wisdom and purchased their small farm a considerable distance from any metropolitan areas. The price had been right. Though a tough go at the start, once they had established themselves as organic farmers and developed markets in the neighboring small towns income improved by degrees.

A difficult decision had come up a few years back when the federal organic certification kicked in and the fees kept going up. It was Mary who had said, "We don't need it, our customers know us and our ways." So they had dropped their 'certified organic' distinction. They just put on their labelling, 'no poisons used, ever'.

In the winter, when time allowed, the family worked together on what had started as Dan's passion, wooden toys. He loved to invent shapes and movements for toys starting with a block of wood and a carving knife. What had began as a diversion, a hobby, had slowly morphed into a family project with everyone taking some part. Hard to say when it started but someone had passed on the information that the Curlew family made these incredible toys which they just lined up on a high running shelf around their dining room walls. Before long people were purchasing the toys and ordering more. It became apparent that Dan and family could be doing this full-time if they wished. As these thoughts passed through their heads a representative of a toy company found them and offered substantial money for the toys, the designs, and Dan's time. When Dan responded, 'when would I have time to farm?' the representative laughed and said 'don't you get it man this is your ticket off the farm?' Dan just shook his head and said no thank you.

These days they still make toys, only difference is that they now enjoy researching about needy kids in their area and leaving the toys (and some farm raised food) wrapped and setting on unsuspecting doorsteps during the holidays. Calley has taken to keeping a photo album of pictures of each and every toy and sometimes they sit together on a winter evening and thumb through the pictures with tears in their eyes. Dan says, 'not ashamed to say it makes me feel good, almost as good as this here farm does."

On a recent home visit, their son Chip was taking stock of his life. Dan and Mary had done well enough with their small farm that they had been able to contribute to Chip's college education. Now, after two years of teaching, he was wondering about the true value of his cow college learning and feeling a strong pull to return to the real farming life he had experienced growing up. Plus, there was a young woman in his life and they were talking of marriage and a family. Calley was listening very closely to all of this. Sixteen years-old and thinking about her own future she was more than a little worried about telling her folks that college wasn't for her. She wanted to study photography, train horses, have a farm of her own some day. She kept it to herself but her feeling was that college was a nasty place of questionable value to her in her pursuits and interests.

A short while after Chip's visit Dan learned that a neighboring forty acres of neglected farmland was to come up for sale. It was only three quarters of a mile down the road from their place. Late one evening Dan and Mary were talking in the kitchen. They thought Calley was asleep.

"I know honey, but that money we had set aside for Calley's education."
"Well, maybe by the time she's ready, in a couple of years, we could make it up?" offered Dan.
"Don't count on that."
"I'm not saying we give it to him. I say we loan it to him. We could even set it up so that he makes the payments directly to Calley."
She looked at her husband intensely. He looked down at his hands.
"You want him to have what you have, don't you?"
"Is there anything wrong with that? I want them both to have what we have, but only if that's what they want."
She laughed.
"I can see it now. I'll be competing at the markets with my own children."

"You always say you've got more customers for our stuff than we want to grow for."

"That's for sure. But I think we owe it to Calley to get her in on this conversation. I'm not against the idea. I just want to be fair."

Calley walked into the conversation she'd been overhearing.

"I think its a great idea. I'm all for it." Then she spoke of her own dreams and reservations.

Dan watched his daughter as she spoke and he thanked his lucky stars for the farm that had given him and given him and given him. And just as that idea of a portable greenhouse had given him a way of seeing possible connections, this moment with his family had allowed him to dream about new layers of fertility and continuity. He smiled thinking about helping his daughter and son with their dreams and how they in turn would help their children. And he knew he was a most fortunate man.

FROSTING IN THE FORMULA

Their farm was small by most standards but it was large enough for them. And, just as important, it was large enough for us, all of us. Instead of an ever dwindling and weakening landscape of a few thousand monolithic corporate farms we are better served by millions of small, diverse, independent fertile family farms. And the more farms we have the greater the opportunity of success for each. The landscape would heal, the countryside would welcome the return of vibrant small farm communities, the economy would strengthen, our capacity to feed people would increase in quantity and health, the immune systems of an ever growing number of people would improve, the governments would move offshore, the moral base would once again rise up from the truths of actual working, and the ranks of the hungry would shrink day by day.

I scratch my head and I think and wonder but I cannot come up with any social model to compete with the small farm. It is our hope. So we must keep saying it, working towards it, embracing it, keeping it alive. The small farm is the answer.

It is the steady drip which eventually drills through the rock. We need to maintain the steady drip. We must never give up working to drill through that rock. In our case, for the true farmer, the drip is more like an incessant drum beat; small, soft and in the background. And the rock is harder to describe or define: it may be the mistaken social notion of what constitutues correct order.

I suspect most people have no collective notion of what constitutues correct order these days. Religions cannot agree. Judges cannot agree. Government cannot govern. The marketplace has become the handmaiden of a corporate communism. And they search for a new pill or gizmo to get us through the day.

But those in the fields, barns, sheds, woods, they know there is a correct order. They know that independent human-scaled enterprise, most specifically the family farms, when gathered together side by side offer the best, lasting, fertile, regenerative hope for the planet and for humankind.

eight

spinning ladders

In the dream, I was climbing a wooden ladder of my own construction. Carving and inserting rungs into the slow-growing parallel upright poles. I drilled the holes for those rungs with my thumbs. I chewed the ends of the rungs to create perfect tenons. The background music alternated between the bleating moans of a large old dog and the sweet and sour trumpet-like urgings of grasses and legumes working their way up through the wet and dry earth. Out around me, some distance away, I could see others building and climbing their own unique ladders of various height. They would not see me, they would not hear me. I had a growing sense that my ladder was near completion and sure enough it tipped to one leg and started its slow elegant spin. It was spinning in the wind of all my hopes, realized and not. It was then that I became aware of an urgency I had worked so hard and so long to deny. Looking across that landscape of ladders, I saw some growing and some spinning and many falling over with their builders, only to disappear completely. But I also saw a few in clusters, standing and waving rather than spinning. I wanted my ladder to be in a cluster, I wanted my ladder to wave into the future, but I had waited too long.

I suppose you could apply that dream in many ways but at this time of my life I cannot help but think of it as a subconscious concern for the future of my efforts. Some might see this dream as the lament of a hermit who wishes to rejoin society. Others might choose to think of this as a man's reflection on his legacy, but I don't feel either. I feel a genuine concern for the ongoing, well past me, and past any signature I may leave behind many years from now.

The main portion of our ranch sits in a remote rock-rimmed wide canyon the indigenous people referred to as the valley of the stand-and-look-at-you-deer. It is sacred burial ground for at least one native nation. It is the home to the largest mule deer winter habitat in the world, home to elk, cougar, eagle, turkey, kestrel, redtail, chukars, and summer home to swallows. Surrounded by forest service land, we are on the lip of the Metolius Basin. It is the sage, pine and juniper-scented, water and feeding center for a vast habitat extending from the Deschutes River canyon to the Mt. Jefferson Wilderness area to the rapidly encroaching developments of the neighboring towns to the south and west. And it is where we try to farm. As one of the last remaining small family ranches in the area, the pressure is on, through taxation and market urgencies, to subdivide and develop. We don't want that to ever happen. Not just during our tenure, but far into the future. We look for the best steps we might take to find a legal designation which will give the area quasi-public, sacred, park-like protection. We don't seek to protect it for ourselves, or even for the general public, we seek to protect it for itself. This effort is like one of those ladders which would stand, spin and wave long into the future, long after we are gone.

Farmlands must first be protected individually by individuals. To exempt lands from categories of use through government edict only protects them for future abuse. The best way to protect farmland is to make the highest and best use of that land to be farming and to allow the extended community to embrace it and her farmers as sacred to their landscape. You do that as a farmer, as a farm family, as the farm's community. You do it locally and you do it with religious fervor.

I am reminded that we hold within ourselves so many secrets of the doing here which we take for granted. We know where the water is, where the sun warms the ground first, where the grass is the sweetest, where the elk bed down, and how to read our sky. We remember what crops have failed and which have bloomed. We fondly recall those of our efforts which left the stamp of greatest beauty on the place. When we are gone, what will happen to those secrets? Once again, I believe, unlike the dream, that our time is far from up. I believe we should be working to share all of those secrets with others who might be dedicated to their own overlapping futures with these doings and this place.

I once read a quotation of Edward Curtis, the famous photographer of the North American Indian, which went something like this, "As each of these old people die another piece of their culture dies with them." His observation, his lament, came in the context of a time when those cultures were, in many cases actually outlawed and certainly dwindling, so the death of an old person likely meant that traditions would also die. For three decades now we have seen the same thing within the community and traditions of true farming. An entire culture which embraced, practiced and worshipped a farming trust, a way of working which succeeded, and worked better every year, was attacked at every social and economic level. So with the passing of each old farmer, and each old farmer's farm, we lost pieces of our culture. We lost ways of listening, working, smelling, touching, measuring, tasting, feeling, valuing, and loving. That critically vital way of life is not completely gone yet, each of us holds on to a small corner of it and we have done tremendous good work to keep it alive. But now we must add a broader level of concern. We must attribute to ourselves that same preciousness we once attributed to those mentors we lost. We have to go into each day asking what it will be like when WE are gone. Now we are the torchbearers. And we need to train and support those who will take the torches from us because what we have saved and built upon is more important than we are. As farmers our cause has been the cause of the peaceable kingdom. I have written before of how it is that I believe the small independent farm is the answer to our environmental woes. Perhaps I should have been adding, all along, that it is also the answer to hunger, deprivation and war. If our peaceable kingdom is to continue to flourish in our small corners, and then move out and into the wider world, we need to see it as our duty to think about passing on what we know, what we care about and what we have built.

There are those amongst us who have been fortunate to have 'inherited' their farm and/or their farming. Not just the piece of land with buildings and fences but more or less a way of life and a living fabric of traditions. They have a tangible sense of what it means to be carriers of that tradition. Many of us made the farm and the farming for ourselves from scratch. We have a far less tangible sense of things going back before us. We lean on instruments such as this publication (Small Farmer's Journal) to give us connection to a wider remembrance and solemnity and to give us permission to be thankful for our work; give us permission because we live in a society and time which honors tradition only in so far as it might be marketed. All of us in this far flung community, this sub-culture of small farmers and ranchers feel some measure of regret that what we have built, and are building, feels fragile and transitory; feels like our own short lives, when we suspect deep from within a collective genetic memory that it needn't be that way.

Each of us will pass on - or pass away. We do have a choice. In this regard, you don't hear that very often these days. <u>We have a choice</u> whether to pass on or pass away. I am not speaking in a biblical sense. I am speaking in the sense of working traditions. It has taken me a lifetime to arrive at certain precious working conclusions about cattle, horses, soils and procedures. Over a short quarter century I have come, with my volcanic passions unchecked, to fear, respect and love this piece of ground we have been entrusted with, our little ranch. I want, in the most complete sense of the meaning, to pass these valuable things, this culture, on. Not on a given Wednesday, when I might board a year-round cruise ship in permanent retirement. (Something I will never do.) You die off by passing away. You live on by passing on. I want to pass the culture of my life on slowly, over the ripening time of my best years.

I woke one morning thinking "I am one who remembers what it was like in my beginning." Then I jabbed myself with a "Mister, most everyone remembers what it was like in their beginning. There's nothing special about that." Fact is I may not be unique in any regard. My story is hardly worth repeating except in its ordinariness. Yes, I may have some gift for stringing words along or for drawing or even for horses, but those are not stories that necessarily link me with you. What links me with you are those ways we meet in our experience. What I mean when I say my story is ordinary is that it is composed of wants and wishes, struggles and triumphs just as yours is. We share enthusiasms where we share enthusiasms. We share frustrations where the frustrations are. I might tell you amazing stories of my time with senators and congressmen, but it would bore you because we don't share that world. But if I were to say that I have a horse who is a booger to trim and that I have figured out how to get it done, you are all ears. You are interested in my beginning and I in yours. You might be interested in my today. I certainly am interested in your today. My today is a new place for me.

We have built a tall, beautiful and vital ladder which needs a cluster to assure it will stand for the longest time. We are looking for that cluster.

And so it is also with our ranch. We are looking for that cluster. I am determined to have our farming and our land be two of those tall, slowly-spinning ladders that stand glowing for a very long time.

This is a new place for me. I feel added dimension to many of my choices and valuations. I liken it, in my own visual thinking, to a relay race. The only difference is that coming up from behind me and all on my team are several runners I am coaxing, ready to receive the several torches I bear. And as I hand them off, at different intervals, I plan to continue running alongside for a 'far piece', being part of the cluster, jumping, as I laugh, from one spinning ladder to the next.

nine

the larger view is the smaller view

Inside the Circle

We reach out beyond ourselves in pursuit of success when the answer might come from understanding our own immediate circle, community, or small world and reaching within it.

Over a quarter of a century ago I met up with a dramatic example of this in Maine. Parker Sanborn was a dairyman who found for himself a way to truly own the vocation he so passionately loved, this at a time when most dairymen were, themselves, owned by the work. Sanborn kept Jerseys because he loved them. And he milked them all in a synchronized season which gave the family months of holiday each year while the dry cows rested during the last months of their pregnancy. 'Seasonal dairying' is what it came to be called. Parker also raised his beloved Percheron horses employing them with regularity at the task of spreading manure on ever improving pastures. He made no hay, contracting with his neighbors to purchase theirs. Parker raised cattle, horses, and pasture while producing milk in the simplest of circular farming worlds. He came to this with deliberation, understanding that what he wanted was a sustainable regenerative comfort in and with the working world of his choice. For him profit in any traditional sense was after-the-fact, almost as if a waste product. His life and its successes were all about the 'top line'. And that paradoxically gave him greater profitability and viability. It wasn't about how much his gross income was, it was about how much he kept and how well the work kept him.

The normally long-lived Jersey cows were even more so for Parker because they weren't pushed to grain-heightened year-round lactations. Parker fed no grain. The cows produced their milk from lush pastures, and in the winter they were dry, resting, and enjoying hay.

Sanborn owned a manure spreader and very few other pieces of machinery. He had no equipment mortgage, no large fuel bills. His costs were at the barest of minimums and his revenue was from milk and horses. He owned his working life. It was a treasure for him.

Many, if not most, of the people of the world work to live. They do things in exchange for money to house, clothe and feed themselves and their families. Frequently the work they perform is distasteful, harmful and/or boring to them. They do the work in order to live.

A lesser percentage of the people in this world live to work. They love who they are and what they do so much that they don't want to be separated from it. In some cases this work provides physical sustenance for them and in some instances it does not.

That first category of people struggle with goals for success. Above and beyond sustenance, if they have the luxury, they work hard to get someplace other than where they are. Their daily work is further clouded by wishing they were not there, by wishing they were elsewhere. They are seldom grateful.

That second category of folks often enjoy their many opportunities to be fully aware of how fortunate they are. Fortunate to be working at what they love. But even they have periods of longing or wishing or planning for something next, something more.

Hard work is essential but it's not the answer.

It is possible to have a goal yet never really see it, never visualize it. Having a vision means in no small part being able to see what is not yet formed. We can have a fixed goal and work terribly hard towards that goal without ever getting there. It is heresy to say such a thing in this country, where the 'American Dream' is sacrosanct. It is hammered into us throughout our lives that if we just work hard all good things will come to us. Fact is, useful as that adage may be there ain't much truth to it, it's a haired-over candied myth.

There are scads of wealthy people who have worked and do work very little. And, inside and outside of farming, there are billions of good people who have worked exceedingly hard for lifetimes without realizing any store-bought notions of success and comfort. There is no simple formula, no one way, no direct shot to success. 'Hard work' alone does not make of us a guaranteed success any more than success defines us as selfish. Other elements must join the hard work, elements intrinsic to each individual's character.

One of those is 'vision' or the capacity to see what might be possible or at least imagine what we might want to make real through our efforts. You can work harder than the dickens to build a fence but without the vision to see what might be when you are done that fence work may be worth a little less to you. When we have a goal or destination in mind and the vision to see it all, the work becomes more voyage than effort.

Another of those elements is perspective, or the ability to back up and take a gander at what is and has been, the ability to see the parts that make up the whole, the capacity to measure intangible things within the working life. It is perspective which gives us a chance to move our goal posts without changing our goals. It is perspective which magnetizes and amplifies gratitude.

We've said it here before and the specific reality of this day and time in our collective history just increases the truth of it, 'there has never been a better time to be a farmer'. What of the obstacles, you might ask? Most of the obstacles exist within a particular perspective. Change the perspective and some of those obstacles become opportunities. The industrial perspective is tied to the linear view, we look 'out there' a long ways towards ever bigger operations, ever greater gross production numbers, ever bigger dollar amounts. That perspective then haunts our beginning choices by dictating the nature of our immediate involvement. We think we need a certain amount of land of a certain quality and location in order to have a legitimate beginning with farming. We think we need lots of money to get started for that land purchase as well as for the machinery, chemicals, livestock and such. What a bunch of hooey! May we suggest that what is needed is not the long view but rather the short view?

Craziness and the Jugular Vein of Society's Best Future

In my interpretation of the story of the Tower of Babylon the people of the ancient world were all of a language and culture and they worked together to build a tower to heaven. It was a huge undertaking and in the process the commonality of the people became confused and started to fall apart with many new languages and subcultures sprouting up. The project was never completed and the people split into factions, nations and religions and went their separate ways with lots of consternation. I see some important parallels between this immediate time in human history and that parable. What follows are a few examples of what I mean, each of which poses an opportunity and a threat to good farming and the independent family farm.

The so-called global marketplace is in for destructive and invigorating challenges.

1.) The market rush to synthetic fuels has handed a windfall to oil crop farmers worldwide. Corn, soybeans, switch grass and oil seed crops are being planted on every available acre as mid to large scale farmers are quickly lining up to take full advantage of the booming market. Meanwhile the traditional uses of these crops will be challenged. Livestock and human feeds are rapidly taking second fiddle to the "futures market" flame and broil. Soil conservation be damned and happy days for genetic engineering as our "war time" economy screams for unsustainable productivity, the nature of which will guarantee accelerated environmental degradation and climate change. They call it a rush towards energy self-reliance but it ain't.

There are good things about this for the small farmer as the traditional agricultural community will be desperate for local food and feed production. Translation: prices for all locally available farm commodities will go up up up.

2.) A few billion Chinese have discovered they have a taste for dairy products and new income to satisfy that taste. Add to this that the North American diet fashion is shifting and milk is once again becoming a sought-after beverage. European agriculture is tied up in the court of tradition and relativity with no milk production increase in the future, while Asia, Africa and South America each have more important things to think about. Result is that for the next decade milk prices will continue to chase fuel prices with no hope of stability because fuel gets the corn. So the cost of precious industrial grain-based milk production will only go up faster. Right up until the Chinese, Brazilians and New Zealanders discover corn, Holstein cattle and carousel milking parlors (those monstrous motorized systems which allow hundreds if not thousands of cows to be milked continuously each day). At that time in the near future we can expect another collapse of the American Dairy industry. But before that, today we have the tragically perfect market environment for run amok genetic engineering and growth hormones (read BHT). Not to mention National Animal Identification Systems.

There are good things about this. It means that local pressures will grow to allow small farms to milk ten or less head and sell the milk direct and raw. Conceivably this could mean 10 to 15 thousand of today's dollars per month in revenue for the small dairy. Using the Sanborn seasonal grass-based system and a ten month lactation with Jerseys, or Ayrshires, or Guernseys, or Milking Shorthorns, or Brown Swiss (all potential long-lived dairy breeds) a small independent family farm of limited but well-managed acreage could pull in an excess of 100,000 dollars per year in milk sales alone, while contributing to the health and welfare of its immediate community.

3.) In today's North American food marketplace, diet and fashion race to the 'chic' organic equation. All externally available numbers point towards phenomenal economic growth in the so-tagged health food sector. Inevitably, for this century, this is leading towards organic industry consolidation in retail, distribution and processing sectors. And that consolidation screams for some additional flexibility in the already suspect national organic standards

For we individual small farmers it can mean a "temporary" windfall or rising prices. In some direct markets $10+ a gallon for true organic raw milk (translating to over $100 per cwt), $5.50 lb for pasture-raised broilers ($30+ per bird), .50 per egg, $4 to $5 per lb for ordinary organic produce, $250 + per ton for premium organic hay. Now all of a sudden many small independent organic family farms are looking at profitable returns of several thousands of dollars per acre.

4.) Billions of people are throwing their scattered ideas, scams and phobias at the world wide web resulting in intensified social polarizations and the wholesale degeneration of knowledge feeding unrest between groups, sects and nations.

Precious nuggets of possibility and probability trickle out of the internet helping to speed innovations in appropriate technology and interconnectedness. Small farmers are finding that the web helps them to make sales.

Drawing the circle

But these good-sounding opportunities are all temporary and terribly unstable. And that is especially so if it is our requisite status. If we insist that this is why we are farming, for these sorts of economic returns and more, then all is lost. If we take the long view rather than the near view, paradoxically, our future is in jeopardy. If we take our own personal version of the short view such as the Sanborn example, our success will be something we can see the perimeters of.

Much is being made these days of the supremacy of locally grown foods and these are good concerns and directions. It is true that the consumer has a far better chance of securing food safety, quality and freshness from local suppliers he or she can know and trust. The flip side of the coin is also true. The farmer who cultivates local customers has a far better chance of security, satisfaction and happiness. Imagine the concept, appreciative customers you can know and trust. Customers who have an emotional investment in the success of your, 'their' farm.

It starts with drawing an imagined circle around your farm, community, family and goals. And then travelling to the edges of that circle and looking within for what it is you want. Each time this has happened, regardless of whether the farming venture is on the fringe of a city or deep in the outback, the farmer has discovered far more customers than he or she could reasonably grow for. Often the circle needs to be drawn in even further to meet the scale of the farming adventure. The short view gives us the best chance for right livelihood and a heady symphonic personal success. But we need to be looking for it with a view towards hanging on because this useful perspective can be a slippery thing.

Society is in a world of hurt. The planet being but one of the stakes of the game. We need to believe in what works, what might bring us back to fertility, manners, and health. In other words, we need to believe in a best future. The small independent farm working within its own circle is what works.

After a long, sustained summer rain, the new fleeting light comes into our landscape as though filtered through a wet windblown prism. It is a fresh steaming luminance, bright and useful. Perhaps that usefulness is why we don't want it to go away. We want to hold on to it. It often puts us in a meditative state as we look cocking our heads sideways and wondering how it is we never noticed that tree before or how the colors on that hillside work so well together or how the space between that distant hill and this rock outcropping has a definitive shape.

Our forest, when wet, goes to the darkest range of color fighting off the most penetrating of that bright fresh light. All about contrast; we see that darkness in the way the light contrasts with luster. And we rest ourselves in this perceptive bath seldom noticing how illusive it all is. It passes without trumpets. And we are back to previous familiarities. But back to that moment back there, that moment when a twist, a shift, a rainfall had given us a new perspective, what was done to our inner and outer selves? Can we learn anything from that brief moment of wonder and gratefulness? Yes, we can.

It's not about our landscape of forest, field, hope, vision or needs. It's about how we walk through those things, it's about the inside of our circle.

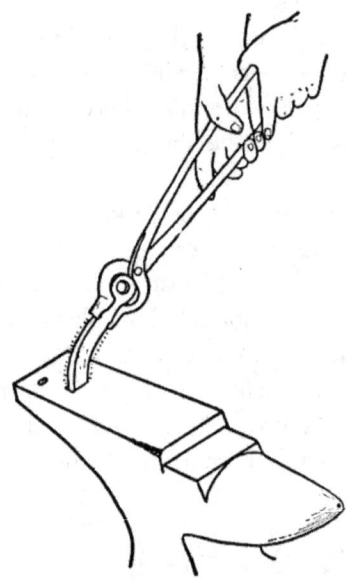

ten

cultivating the family farm

The poetry of each motivation holds the seed.

Love hurts but justice is not enough.

What is it we do? Are we farmers or part of the broader farm support community? Or something altogether else?

How do we do what we do? Are we organic farmers? Are we holistic in our approaches? Are we industrial farmers? What morality informs our methods? Do we care?

Why do we do what we do? This is the most important question of all and it informs our best options for the successful cultivation of the small family farm. The poetry of each motivation holds the seed.

But I left something out. That would be the introductory question. Who are we? Who are you? What makes us tick? You may see how it is that what, how and why would answer much of that.

Who are we, you and I, and what morality informs such judgments?

Labeling has become so important to us as farmers. Not just the question of 'certification' and the listing of contents but also the clarification and acknowledgement of our place in the world. The very definitions of our work and lives.

There are piano wires strung tightly across our path forward, at eye-level, throat-level, ankle-level, criss-crossing in a deadly, barely visible web. We collectively helped to string those wires, we handed authority to the USDA and to NGOs and to industry in a blind stupor of gullibility and now we have Hansel and Gretel's forest to traverse and the witch's ovens to avoid. But none of it is as bad as it seems for the simple reason that much of it disappears with non-compliance. It is that perfect and hideous nightmare which disappears if you come awake. We still have the right to choose which party we go to and with whom.

At the beginning I asked what, how, and why, informing who. I said WHY is the most important thing of all. Now to confuse matters I want to interject that WHY may owe its ultimate value to how much.

The Hand Off
In the summer of 1986 I was visited on my farm during haymaking by Tom Forrester, two translators and an eighty year-old Japanese dignitary named Tedeo Ichiraku. This man had been the Japanese equivalent of secretary of Agriculture following WWII and is generally credited with industrializing and modernizing food production in that country. Following that term he had an epiphany: he regretted all those things he had set in motion. He saw them as destroying the very culture of centuries old farming in Japan. He wanted to do something to correct what he saw to be the mistakes of his administration. His answer was to form the successful and important organic farming organization in Japan. It was around 1980 that he discovered *Small Farmer's Journal* and became a reader. In the mid eighties, eighty years old himself, he saw the end of his life fast approaching and he felt it time to hand off his torch, so to speak. When asked to do a presentation to the organic farming folks in California he had Tom Forrester set up a visit to my farm in Oregon. He thought perhaps I would be one to understand and use his message. And he wanted to see for himself an application of true horse power.

On the day of his visit I was unloading loose hay from a wagon by grapple fork and trolley. Tip, our Belgian gelding, was pulling the haul-back rope up the back of the old, large, gambrel-roofed barn. When Ichiraku's entourage arrived he instantly, in expensive silk suit, trotted to the barn to watch the procedures. Quickly realizing what was involved he informed me, through an interpreter, that he was going up into the hay loft to watch the load come in and dump. My rejection of the idea on the grounds of liability and danger fell on deaf ears. The elder statesman, with assistant, climbed the wooden ladder. When the load came up and through the mow door, eclipsing the light, I heard him cheer in Japanese. As the load rolled along the trolley to the dumping point, his giggles got progressively louder until he shouted something jubilant. When he came down his suit and hair were covered with pieces of hay framing a radiant smile. Later in the house, through his interpreter and Tom, Tedeo told me his life story concluding with these words. "Soon I must die and it is important that I pass to you this discovery which has come to me too late. Organic farming alone is not the answer. When it becomes profitable the big companies will do it and destroy it. The answer is scale. We must do our farming small as the size of a man, as the size of his family. In this way it will belong to each and every one and it will be healthy and strong. You must carry this message, you must tell everyone. Thank you. And thank you for showing me your work. I love your great horse and the hay falling from the sky of the barn. Your farming is good, it is the size of a man."

You On Your Own
Scale is ultimately about singularity, it is about embrace, it is about what we can hold, what we can touch, what we may know intimately, what we are able to fully honor, what would know us as important part. It is about being able to laugh at ourselves because we are accessible to ourselves.

We are a force. Not together so much as singularly. We, each of us, are a force.

There is no denying that in the aggregate we are something altogether 'other'. We must improve on that 'other' without detracting from the critically important singular.

Your dreams belong to you.
Your plans belong to you as do your operations.
Your successes and failures are your own.
Your frustrations, fears and opportunities fit only you.
Your satisfactions and sense of self worth shape you.
You are singular.

But now, in this crazy and shifting time of environmental, political and economic cataclysm, the collective force and essential security afforded the world by the community of small independent farms is important beyond even the wildest measure. Small farms feed the world. Without small farms the world is no longer fed and the consequences are catastrophic.

We know we must save small farms, preserve them. But both of those measures, saving and preserving, feel all the world like cures to a problem. I think we, instead, must talk and wonder and act about conserving small farms. That approach, conservation, feels more akin to prevention - preparation - parenting. It feels more in tune with growth and life.

We cannot talk about the conservation of the family farm without speaking of the singularity of the farmers. Just as with seed saving, the preservation of the small family farm cannot be done 'on the shelf' or in museums nor on living history farms. The institution of the family farm and the community it engenders needs to be 'cultivated' in order that it live on. And by cultivation we mean all the work we do from the planting of the seed or the birth of the farm all the way through to harvest of the crop or the transfer of the farm to the next generation. Unless seeds are planted and successfully grown it is only preservation, there is no conservation. The farming and the land are conserved when the farmer protects his or her effectiveness. And when we speak of farming and farmers we know that fertility and effectiveness share aspect. Motivation is a form of fertility. It fertilizes our effectiveness. We put our future as farmers in peril when we deny the elegance of our motivations. That elegance stems from the rhythm and architecture of our enchantment. It's about magic carefully held, instinctively used, and always revered.

The Blue Corn

A quarter century ago or so I spoke at a Land Stewardship Conference in Colorado. It was there that I met John Kimmey of the Talavaya Center. John told me a story which has stuck with me in haunting ways.

John, as anthropolgist, worked with a native elder in the Four Corners area of the American southwest. He was keen to help to save the agricultural heritage of the indigenous people. The native elder talked to Kimmey of his concern that the saved seed had sat too long on his shelves. The young people did not care. He worried about the strength and vitality of the seed. After gaining the elder's confidence, John was given a jar of a fading blue corn seed. The old man had taught John about planting sticks, planting rattles and about the songs to sing when growing corn and beans. He told John to plant half of the seed in a plot of concentric circles. The other half of the seed was to be planted "out of earshot" of the circle plantation. Kimmey was to plant the second portion in a traditional grid of parallel rows. The circle was to be done with planting stick, rattle and the appropriate songs at the right times. The traditional rectangle of parallel rows required no special treatment or ritual. Following the instructions to a "T," John was amazed at the difference between the two plantations. The circle grew strong and luxurious while the rectangle looked ordinary. And when harvest time came, the circle produced vibrant blue seed quite different from the yellow seed of the straight rows. John took the results to the elder, excited to share the outcome. The elder asked Kimmey what the bright blue seed meant and John said he wasn't sure. The elder hit John with the planting stick and, tearful, he said "Don't you see? The seed has remembered the songs."

Completing the Definition

You as singular farmer, as singular would-be farmer, you belong to us as part of our collective definition, part of our collective songbook. And that definition is a work in progress, yet to be completed. We must take the steps to complete the definition, and write the last songs, before we lose what we have worked so hard to build.

We have this arcane thing society has identified as intrinsically valuable, we call it the small family farm, others call it by different names. Over the last almost four decades we together and individually have worked to give new aspect and vitality to our working understanding of what a humane small family farm is and can be. In our midst are examples of exciting fertility and economic vitality as the result of the intoxicating, almost erotic, blend of the best traditional methods with whole new approaches. But with new successes come the pressures which threaten to dehumanize, to desensitize, to deaden this evolving new farm. Though it may be predictable that significant successes with new approaches would move us towards industrial models, it is NOT inevitable.

People ask me 'how is it you get so much done' when what they want to ask is 'what keeps you going?'

I try, whenever my morning suggests a tedium or predictability, to sit still and concentrate on a simple mantra. I admonish myself to "pay close attention, very close attention to this day." That little exercise seems to give me back my enthusiasms and entice me to think wider about what's next. But, even for me after all these years, I will confess to you that on some particularly unfortunate and moronic days I get down and wonder at my own ineptitude.

I realize that for some of us it has become about watching ourselves get old, watching the work we loved get old… It doesn't have to be that way. We must work to avoid this.

You are an instrument which must not be wasted. The key lies within your motivations. The more elegant they are the greater your energy. When I say you must not be wasted I do not mean that you need to work harder. I use the word wasted in the vernacular as in burnt-out. You are a vital instrument which must not be burnt-out. There was an elegance in those reasons, wishes and visions which brought you originally to this work. When you allow yourself to get 'wasted' that elegance disappears and your motivation wanes.

When you feel yourself at risk of burning out (or long before) you MUST take time off to rest and do other things.

Keeping it Altogether
The poetics of an elegant motivation include Insurance, Assurance, and Relief, three things which can be difficult or impossible to purchase. But they are things which we might gather unto us together.

I have for decades trained horses for work, I have learned that the best animals are those who are worked regularly and who are always in training. Each time I allow an infraction of my working standards of conduct I am telling the animals that the rules don't count. This makes of them lesser work mates. I mess them up with my laziness. When I discipline myself to constantly reinforce the behavior rules for the horses all of us are better.

The same thing may be said of us as farmers, only we are in charge of our own training. Each time we let rules of behavior or work slide we become less effective. Tedium will tempt us to cut corners, to court the more simply predictable at the expense of the magic that did and will and could motivate us.

Because cultivating the family farm is about conserving this model unit, we all need these 'units' to do more than survive, we need them to thrive. The difference between surviving and thriving is like the difference between subsistence and self-sufficiency. Distilling further, it's not about family farms plural, it's about these in the singular, each one. And further yet it's all about the individual. What do you want, what do I want, what do we want?

The very nature of the farming endeavor would seem to suggest individual connections with growing, nurturing, and creating. Though some might not think of it in this way, we are drawn to this involvement with the natural world because it just might make us feel as a worthy human. We feed the world, the world needs fed, we feel good about what we do so we continue, the system continues (in spite of official efforts to the contrary). All of it depends on us feeling good about what we do. If that isn't there the rest crumbles and we no longer have small family farms. So, just what would make us feel good?

Is it money? Is it success? Some of us have fallen repeatedly for the bottom line trap. We come to believe that if we could just get our sales up to X all would fall in place. While there is much work to do and tremendous potential within the questions of pricing and marketing we are at risk of accepting the inevitability of the industrial model. "If only I could get a quicker way to load a thousand more bags of carrots?" "At $5 per pound we're selling every broiler and we have a waiting list. And that's with 100 six pound birds. We could easily sell 500 birds. But there has to be a more efficient way to slaughter these birds?" Sometimes we get so wrapped up in the posture of the perfect dive into the pool of growth that we forget to check to see if the pool has any water in it. It's like bungee jumping with a cord that is a foot too long.

Mixing it Up
We need diversity in our lives as much as in our farming. Without it the tedium of process sets in and destroys us in turn destroying the fertility of the farming and of our dreams.

Forgive me from quoting an excerpt from my earlier book *Farmer Pirates and Dancing Cows*:

"We are told we must narrow our focus to a particular vocation or discipline and work hard on that one thing otherwise we will diminish our chances for success. The man who would be a cabinet maker and a botanist is asking for trouble. The woman who would be a dancer and a cellist is doomed. The child who would pursue automotive engineering and poetry is labeled a self-destructive curiosity. The best and most dynamic individual examples of diverse enterprise are frequently hidden from full view as an act of self-preservation. We succumb to the pressures of family and community and deny the truth of our potential. We follow, short-sighted, the social edicts of our time. The result is often an unhappy life, ironically even when we are immersed in the work of our choice.

"Alexander Borodin was a chemist and a surgeon, but we perhaps know him best as a classical composer of the highest rank. We measure him today by the lasting power and grace of his music. By all accounts he was a happy, fulfilled man. Yet in his day, peers within the communities he worked would have doubtless wondered about his seriousness pointing to his various vocations as indication that he seemed spread thin. I prefer to think of him as spread thick."

I am a painter and a writer and a horseman in concert with my farming. Though often the competition for my limited time is fierce, I refuse to trade off any of those aspects of my working life. They inform one another. For example, with my painting I am always in pursuit of an elusive quality to my work which I identify as changeability. I want to create pictures which are constantly shifting, changing, and challenging the viewer. I want this because I have identified for myself that when a picture, or a piece of music or a farming labor becomes predictably static, it also becomes tedious and eventually even something to avoid. I simply no longer want to spend time with it. And the opposite is true, when a piece of music, a farming labor or a painting is, upon each return, fresh and exhilarating, this permanence to its aspect refreshes me, lifts me, it invigorates me.

That is why, after almost forty years I still choose to depend on draft horses as motive power for the farming. I never tire of them or the nature of my working partnership. It is always fresh and intriguing. In the oft quoted words of my mentor Ray Drongesen, "The longer I work them, the less I know, and the easier it gets." There are many other choices to farming that can have the same refreshing aspect. They can be identified for most any operation. And they need to be utilized to keep the enchantment in our work.

I know I am speaking of something that is not quantifiable, something which is intangible. It doesn't fit anywhere on the balance sheet. But is that necessarily true? If we can feel the force of such things does that not give them shape, form, weight and aspect that we might hold? It has been said that the aspect of which I speak is actually nothing more than a shift in perspective. Alter how you see a thing by moving slightly and for a moment at least your measure of that thing alters as well. "So what?" The economist would say. So EVERYTHING I answer. We are talking about enchantment, a thing we must protect and grow if this small family farm idea is to continue beyond us.

We chose this life of farming. If we chose it because we were attracted to this way of life, then we have more than a little WAX in our nature. And that wax makes us melt in the presence of great, specific sentiment. This is clear indication that the enchantment is close at hand. But that same wax would make us vulnerable to the tedium, stress, terror and melancholy of our taxing moments; when the magnificent crop is destroyed overnight by bug or blight, when the prize livestock dies prematurely, when the pump burns up, when the barn burns down, when the loved one disappears. When of a moment we find ourselves muttering, "I can't take this any longer!", we can feel the wax of our better nature melting. That's when it's time for a break, a holiday, a vacation, even an extended sabbatical. Our North American culture disdains rest and recuperation. That's a mistake that our culture pays for in many ways. With farming no one is telling us, as independent operators, that we have to work 17 hour days, day after day. We tell ourselves this malarkey. We can just as easily tell ourselves that we need time off. And I don't mean a day or two, I mean serious time away from the farming and the farm, relaxing, doing something altogether different, seeing new people and things, following a secondary dream, sharing experiences with family and friends. Revitalizing ourselves. Because we need distance to clarify our vision, we need to be able to back up, way back, and see what we are doing. And that distance can be more than a clarifier, it can be rest pure and simple.

Studs Terkel remembered a taped interview he did years ago with a poor, uneducated but very bright single mother in Chicago. She had never seen a tape recorder before, let alone been recorded. When the interview was over he played the tape back for her to hear. As she listened her eyes got big and her hand went to her mouth. And when the tape was done she turned to

Studs and said "I had no idea I felt that way."

The distance will help us to see.

But when we have seen, we need to move back in close, so close that we find ourselves inside. Because today we need to circle all of our ideas pointing them, nose in, so that they must eventually see one another. That center where the sight lines cross will electrify and feed back to each the tangential possibilities of a nearly infinite number of variations, all on a theme of right livelihood. It's about cultivation as an act of self-preservation and conservation. Abstract yes, but it can be actualized.

Sometimes the impossible utopian view gives shape to workable even useful suspicions. I believe what we lack in this fractured and far flung new farm community is the actualization that would come from a realized sovereignty of our collective identity. What we lack is a physical address to our collective nerve and value core. Imagine with me that somewhere not far from here there was a mixed crop and livestock farm upon which, in one corner there was a dedicated cemetery. In the second corner was a one room school house with adjoining facilities for farming workshops and clinics. Go to the third corner and see a small out-patient clinic, a country hospital if you will, where emergency medical treatment and birthing is looked to. Now in the fourth corner see, with the perpetual backdrop of the working farm, an assisted living facility for senior farmers and their families. See this facility under the umbrella of a membership which might facilitate farm caretaking services for farmers who need sabbaticals. See also that this membership pools its numbers to acquire an insurance package making this commodity affordable and accessible to membership. What we have with this vision is Insurance, Assurance and Relief.

As the Answer

Whether it be with a version of the hypothetical nerve-center farm I described or otherwise, what we are, what we do, the very human scale of our individual efforts, all of us singularly and together, we have what it takes to save this beautiful planet.

I can see it. I can see it in the central valley of Washington and in Vermont, I can see it in the Carolinas, in New Brunswick, in Botswana, in Bolivia, in Chiapas, in Serbia and in Louisiana. And with that vision comes the imagined outcome, the shift in economic priorities, the symphonic sovereignty, the full-on social empowerment, the return to a right etiquette to our interdependence and gratitudes.

Love hurts but justice is not enough.

The poetry of each motivation holds the seed.

Please pay close attention to this day, as who we are may be the thing that saves this beautiful planet.

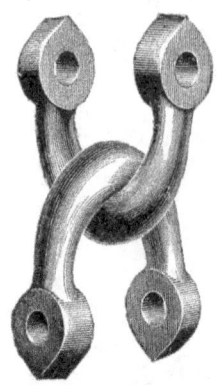

eleven

spun honey and colony collapse

Where have all the bees gone? Many who care deeply are searching. Clues are floating to the surface. I see parallels with the human condition of late. Not to suggest that we live in colonies which are collapsing, but rather that if we did live in colonies or clusters or like-minded swarms we would be better insulated against what seems to be heading our way. We actually live today in a morass of commerce-defined mud which denies us our ability to produce "honey." We have little or no savings or food put by or fuel stockpiled or seed saved. Many of us live without a support group of family and neighbors those who would be ready and anxious to help each other out in times of need. We live day-to-day trusting that what we need will be available and affordable when we need it. We don't produce the sweet stuff of sustenance. We have no "honey" to spin. The means to change this are at hand today. In a very short time we can return to our rightful colonies, clusters or swarms. And we don't have to go backwards to do it.

The new, The healthy, and The hopeful; These things are on my mind.

The new? We must trust that bright young minds will be captivated by the small farm community. It is my personal sense that we should speak to them as equals, never toning or moderating the message just because we may have the edge with experience. Keep the carousel spinning at the same gainful pace. Those young people may be our golden rings to reach for, but I prefer to think they will be attracted to the shapes, colors, possibilities, and landscapes viewed as we spin by, busy in our dedicated usefulness. Rather they should choose to jump aboard and join us.

The healthy? High fuel prices and erratic global market pressures are rapidly reacquainting folks to the beauty, strength and security which may come of local self-reliance. Nearby farmers and artisans are increasingly valued as the cost of shipping goods from around the planet becomes more and more prohibitive. Chairmakers, shoemakers, market gardeners, cheesemakers, dressmakers, woodcutters, blacksmiths, wooliers, candlemakers, et al., supplying inside of 30, 50 and 100 mile circles of community, this is a working definition of healthy.

The hopeful? Within this time we see the challenging economic, political and energy climate actually rewarding new, creative, independent, small organic farmers in highly lucrative and sustainable ways. Not only are they increasingly valued as members of local society, as mentioned above, they are also poised to be extremely well-paid for their work. Add to this that there has never been a better time, in all of recorded history, to be a well-informed and well-equipped horsefarmer. Whole lot of new, healthy and hopeful.

Those of us with tenure on a piece of land, with tools to farm that land, and with the knowledge to farm are amongst the most fortunate alive on the planet today. Yes, these are difficult times. With every passing day the food supply is thrust deeper into uncertainty what with wars, famine, pestilence, weather changes, corn ethanol, fuel prices, banking insolvency and government meddling to name but a few concerns. The agribusiness community, as orchestrated by multinational corporations and the USDA, continues to mess up farming in ways which can only be described as stupidity and shortsightedness feeding greed. The result? Big farming is collapsing in on itself. So the truly independent small farmer, with increasing success supplying his or her own needs while selling direct to local markets, is in the cat bird seat.

Rather than to technological or biological innovation or industrialization or commercialization, the future of agriculture belongs to mastery of the craft. The notion that more food and fiber and better food and fiber will only be produced by those who are fully invested in the natural processes of farming is the big new idea for the future of mankind. Add to this that these self-same directions always return tenfold a better environment and stronger more diversified communities. It is virtually impossible to realize mastery of the craft of farming from the position of large-scale industrialized agriculture. Appropriate human scale is of paramount importance, scale and independence. The truly independent small farmer is the new farmer.

These are uncertain times. Think about those very words for a minute. Aren't all times uncertain? When has there ever been a time which wasn't? Yes, desperation, destruction and death have become the mainstay of mass entertainment served through erstwhile news outlets. That might be part of the reason it feels so much worse now. Yes, there is great growing disparity between the rich and the poor. And today there are people with less money than a year ago. Also, today there are people with more money than a year ago. Today we have former quiet spots around the globe and other regions now torn by war and strife. World governments are being bought and sold every day. The more things change the more they stay the same. Nothing new? In almost every aspect, times are as terribly and comically uncertain as they have always been. Every aspect but three.

Our new problems:
1. The climate is changing and we don't know what that means.
2. Oil is over and very few people seem to get that.
3. The world's food supply is hanging by a thread over a precipice.

These are uncertain times but much of what would worry us is completely out of our control. The things which are within our control, however, are rich with opportunity. We have been saying it now for some time but it bears frequent repetition, there has never been a better time to be a farmer. The loss of affordable oil and the commensurate effect that will have on natural gas and electricity can spell doom or hope. I prefer to see hope. Now, people will be forced to look into alternatives and think of all things as local. That spells a growing customer base for products, services and innovation. People will need their food to come from local sources, they will need their clothing and fuel to come from near, crafts and the arts in general will help to denote region and lend important identity rather than be lost to the vulgar brokers of the momentarily chic.

Once again the world is about to become very large. I mean by this that China, once thought to be a day's plane trip away, will soon rightfully feel like it is ages away. Our cavalier attitude with regards to global access is being tested more and more each day. And that will be good. We are already experiencing the beginnings of exciting local renewal. Whether it be in remote parts of rural America or far flung corners of the globe. It doesn't take a crystal ball to see that the era of cheap global air travel is near an end. The catastrophic growth in the cost of freight will force us to produce basic goods locally, especially food. We will not be able to afford cars, steel, off-season grapes or computers from thousands of miles away. And we will find ourselves asking basic questions at that point. Such as what do we really need? Do we really need new Asian, Italian or German cars? Do we really need off-season Chilean grapes? Do we need Chinese computers? I say no to all three. Do we need to produce more of our own steel and our own food? Absolutely and the sooner the better. As for opportunity; those of us producing product and services for animal power know astounding growth in interest and sales. (As dramatic as this growth has been, it is only the beginning. We have barely lifted off the launch pad.) Think about the phenomenal opportunity available to that first company to produce a stripped down, cheap basic electric automobile (no frills - no gps, no on-board computers, no stereo, no video, no electric windows) along the lines of the infamous Deux Cheveux (Two Horse) of France.

Not long ago social pundits tauted the end of the age of industrialization and the beginning of the age of information. They explained that we didn't need to worry that other countries would be growing our food, making our cars, answering our phones for us, because we would be clicking along happily with high paying jobs serving the information sector. What a colossal sham that is! Microsoft, Yahoo and Google have all just about succeeded in designing systems that will require far less people than initially thought. There just ain't the jobs out there in the so-called "information" sector. By the way "information age" is a euphemism for, or another way of saying, 'dude my computer is smarter than your lame-butt computer.' What we have here is electronic regurgitative crap to the nth degree. We have software technicians working on adding aspect to remote control devices so that when you open your garage door from your computer-laced Porsche your cat box will be cleaned and a soft scent of lavendar sprayed at your entrance portico. While across the planet children are dying of hunger at the rate of one per four seconds. Forgetting the grotesque inequities and the complicity with crimes against humanity, what thankless self-absorbed monster would want to live such a disconnected existence as that Porsche driver? I know he would be useless weeding in my garden, and ultimately of questional microbial value as compost material. "Information Age?" Fantasy Island poison, that's what we have inherited from the Clinton/Bush/Obama consortium in consort with Chinese and Russian organized-crime racing headlong into those other two euphemisms "Global (read gobble) Free Trade" and "One World Order." Nasty, nasty stuff and every bit of it chock full of opportunities for we small farmers.

Recently my friend, the poet Paul Hunter, introduced me to this quote from Robert Blythe
"The world will soon be breaking up into small colonies of the saved." from Blythe's 60's volume "The Light Around The Body." It was the Vietnam War era and he was speaking of and to people clustering together in their shared beliefs.

Apologizing to Blythe, Hunter and poetry in general, I want to take that quotation and shake and shape it slightly to fit where I think we may be heading today,

"Time for the world to break up into small self-sufficient colonies of the new, the healthful and the hopeful."

I recently heard a midwestern commodity farmer speak of the phenomenal potential this year's crop had to either make him rich or destroy him. Growing corn, soybeans, canola and wheat, he needed 25,000 gallons of fuel for the year's tractor work. He had contracted for $3.79 a gallon. And he spoke of his chemical fertilizer bill going from $400 ton to $1100 ton. "If the crop does well and the prices hold up, we'll make a lot of money. If anything goes wrong, we'll be destroyed." Tremendous pressure with lateral lightning bolts mixed in. If he does well, he will be able to pay down the farm mortgage some, replace the pickup truck, and payoff his enormous production credit loan. His corn was all being sold for ethanol even though he felt this was wrong and worried about the world food supply. And the pressure to do well was causing him to reverse certain soil conservation practises he believed in.

Measure that story against the organic horsefarmer who, on his 160 acres, maintains his crop rotations, plants his own seed, does all of his field work with his six home-raised Belgian horses, and measures his purchased inputs in the hundreds of dollars rather than the tens of thousands. This man knows, with current commodity prices and whatever weather mother nature throws at him, that he will be able to paint the barn and house this year and help his son get a start with a farm of his own. A true applied definition of prosperity. And as a bonus he can honor his concern that food would always come before fuel.

Once again our new problems:
1. The climate is changing and we don't know what that means.
2. Oil is over and very few people seem to get that.
3. The world's food supply is hanging by a thread over a precipice.

With climate change comes a whole new set of challenges to farming. From region to region we see unpredictable new patterns of cold, wet, hot, dry, and turbulent. For us in Central Oregon we have experienced nearly constant winds with a very long winter encroaching on spring. Good farmers will have to pay better and more constant attention to these changes to try to at least mitigate some of the more catastrophic effects of the weather changes. Having all your eggs in one basket with say one big field of corn or wheat greatly increases your chances of devastation from the weather. This is the time to diversify crops and spread the risk. This is the time to build into the planning second and third options for the crop. For example, if weather does not permit making hay, haylage or pasturing may be the answer. Identify crops which, in the event of damage, might be fed back through resident livestock. Weather uncertainty is a perfect excuse for reducing your expectations. Those farmers who absolutely MUST get a bumper crop at top price in order to survive are asking for trouble. Nothing is to be gained by putting your head in a hole and insisting that the climate issue doesn't exist. Everything is to be gained by being attentive and creative.

Oil is over. What we mean by this is that those days of cheap oil of ready and steady supply are over. If you must continue farming with a diesel tractor you will need, for your survival, to pay much closer attention to each and every trip across the field. It is a perfect time to pay attention to the laws of diminishing return. Calculations need to be made to determine whether or not your farm is too big and too specialized. Your gross income means nothing if your net income is miniscule. Sophisticated implements and tack are now readily available for the intelligent and resourceful farmer who might be personally suited to farm with horses or mules. Imagine farming 80 to 160 acres (or more) and having the opportunity to keep most all of what you make! Imagine having little or NO fuel bill for the farm! Imagine your soils tilth and fertility improving with each passing year. Imagine your children reconnecting with the farm!

The food supply is in jeopardy. Today the food supply is critically tied to a vast infrastructure of shipping, processing and distribution. Interrupt that chain and the supply is jeopardized. Even school children know that the chain is being interrupted by fuel shortages. Our government and agribusiness have made the unethical choice to put fuel before food. More and more people worldwide aren't getting their food.

As farmers we know what it takes to produce a calorie. We may know quite a bit less about how those calories morph and move through the system. It is my belief that we as farmers need to withdraw, where possible, from the industrial distribution complex and make efforts to simplify how our produce can get to people. I realize that is a tall order and one which will smack to some of revolution but the name of the game is getting food where it is needed and making a living doing it. I suggest that the current system may not work, not now, and even less in the future.

I have in the past been accused of spouting the rhetoric of farm anarchy. My accusers point to my insisting on, for thirty-eight years, speaking of the paramount values of 'independence' and 'scale'. Thinkers on the right and left of farm issues believe that farmers need to see themselves as pieces of the larger puzzle and be willing to subjugate their dreams and needs while complying to the defacto market or social rule of agribusiness in general. They say my cry for independence is a revolutionary cry for lawlessness, for a disrespect of the marketplace, for an attack on the contradictory hierarchies of free trade. The argument is that we need the whole food system in all its complexities, and any threat to that system is a threat to the cost and availability of food. I disagree and know that my small thoughts are no real threat to the status quo. But at the risk of mixing up the argument I will add that today I see the need for "independent" small farmers to stay that way while "belonging" to chosen communities, or colonies, or clusters. Not so much because of that old adage of strength in numbers (as in we against them), but because of the sufficiency factor (as in I need your eggs - you need my blacksmithing skills - she needs these fence posts - and he needs that lamb - all of it local). The small farmer doesn't need the global marketplace. The small farmer needs a small community of his or her own.

Some statisticians claim there are 525 million farms on the planet and that 425 million of those are small farms averaging 2 acres per family. Of course those numbers are up for grabs especially in this country where USDA census conclusions regularly discount or exclude the small farmers. But it is interesting to think about those numbers matched against the world's population of nearly seven billion. And to factor in that we are told there are far less then two million farmers in the U.S.A. Conjecture has it that those 100 million industrial scale farms worldwide feed the affluent 15% of the world's population while the remaining 85% of the world is fed by small farmers. Lots of various inequities in that, but most important is that 15% of the world is fat while a child dies every four seconds of hunger.

But it's all shifting and the inequities are increasingly subject to mother nature and economic forces. Small farmers, in their insular strength, will continue to prosper and their ranks will grow. IF they join forces in satellite clusters of the "saved" and encourage growth within their own ranks the world will enjoy the new, the healthy, and the hopeful in ever growing measure.

twelve

farming for life

Farming may be enough for a lifetime
but is a lifetime enough for farming?

Preparing for difficult times

Now we find out what we're made of. Every step we take, every concern we allow ourselves will define our place in these difficult times. We must protect our families and ourselves without forgetting humanity and nature. How we do that will determine the shape of the new world ahead.

Even before the planet's aggravated weather cycles could finish whuppin' our sorry butts with natural disaster after natural disaster, 'our' economy goes terminally ill. The planet is sick and we made her sick. The corporate dragons who promised to feed and care for us are imploding. The coupons we were told to treat as indicators of our 'wealth' have become worthless. The governments we needed to believe in have become hideously self-serving and stupid. And, irony of ironies, the very skills which have been mocked and denigrated for half a century now turn out to be the only things which can save most of us; skills such as food preservation and gardening, the craft of natural farming and the ability to heat and clothe ourselves.

Is there hope? Of course there is; hope, and real opportunity, but sadly not for those crippled by fear and angered by the loss of net worth. And perhaps most important, not for those who can't 'feel' the change and the threat.

There might be a fancy scientific term for what I know as the 'boiling frogs' syndrome. Scientists tell us that if you put a frog in a pan of cold water and set that pan on a stove burner, that frog will sit there until the water boils and the frog dies. Conversely, if you try to put a healthy frog into a pan of nearly boiling water it will immediately jump out, saving itself.

So-called advanced modern society, commerce-riddled as it has been, requires the 'boiling frogs' syndrome be true of us sheep-like humans. We sit calmly in the water of our society as the temperature goes up gradually. Today that metaphorical 'water' is reaching a boiling point as millions face certain unemployment, hunger, and homelessness. And no effort is made by us to leave that pot of water we know as today's commerce-driven society. How do you leave that society, you ask? By taking charge of your lives, by returning to the basics of a self-sufficient existence, by 're-villaging' into communities of like-minded individuals, by growing some, if not all, of your own food, by rejoining the biological world and demanding of applied science that it truly serve humanity and the planet, by rejecting sadism, gluttony, and ingratitude, by disconnecting from the electronics and chemistry which deaden us.

Comedians and politicians, working with and without script, are still heard, even in these woolly days, deriding any who would even think about growing their own food. That won't last long. We gotta eat. And the party in power better realize that soon because we are only three or four missed meals away from revolution. As fathers and mothers, think about what it would mean to you to watch your own children starve. Now multiply that in the U.S.A. by millions and apply those intense feelings of desperation, helplessness and fear up against the news stories of multi-million dollar pay packages for "hired" executives of corporations and boondoggle bribes for elected officials. (They all are employees, not owners. And they control the decision making processes which have resulted in our economic meltdown and put millions of people out of work.) That, I insist, is a post-modern recipe for a revolution, one which we would be well advised to avoid.

So, I maintain that the concept of families working with their own hands to provide some measure of their own food, shelter and heat ain't funny, it's essential. It used to be called self-sufficiency and early on we discovered that when we had more eggs, milk, potatoes and beans than our family needed we could sell or barter that excess. It was good and we called that farming. Though industry took upon itself to raise large tracts of food, it was less of farming and more of agribusiness. And agribusiness with its heavy metals and toxic chemistry has had absolutely nothing to do with self-sufficiency and less to do with sustainability. Now we see the transition to a fresh form of true farming immersed in self-sufficiency. This good new farming has the power to save each of us and our planet. It will also give us, as a wonderful bonus, true sustainability.

Today, amidst the culture wars and an imploding economy, marketing professionals work continuously to claim ownership of the word *sustainability* for their corporate clients, a claiming race which embraces all the toxic vagaries of fashion and fad. That won't last long either.

The quest for sustainability ahead of self-sufficiency is misguided. Self-sufficiency must come first, and when it does come, sustainability will follow...

> *A few months back, we had returned from an extended business trip and I went immediately to turn the irrigation pump back on. The glorious summer heat held promise of damage to this farmer's hayfield if he let too much time pass without the sprinklers going. High elevation sunlight causes a more rapid transpiration of moisture than down at lower elevations. We are up against the Cascade mountains at 3,000 feet.*
>
> *I opened the gate valve from the irrigation lagoon to coax water to prime the pump and mainline. Then I pressed the electric pump switch. What resulted was a terrible, irregular screaming noise I recognized as steel against steel set to in a losing battle. I turned the pump off. Not good. What did this mean? Would I need to replace that big pump and 3 phase 25 HP motor? It would be a bad time to do that. Summer is always tight with so many things demanding what few funds are available. So I calmed myself down and set to thinking about the best way to proceed. I don't like it when things feel insecure.*

What does it mean to be secure? One of the hardest, flintiest first lessons of life is that there are few, if any guarantees. We don't know what weather, money, equipment, family, friends, strangers, health and life herself will bring to us tomorrow. With or without experience of horrible and seemingly inexcusable loss, some people are terrified by the vagaries of life. Some others of us know the comfort of believing in whatever the outcome may be. They have a 'faith' both generic and specific. But even with faith we still want for security in the immediate sense; we want to know we can care for ourselves and our own on into the future.

Even in our modern form, luckily most humans still have a little of the proverbial ant in them. We, the many, busy ourselves to gather up the 'necessaries' to get through the winter, get through the lean times. We gather and store. Some, though, feel the urge yet DON'T follow it because modern convenience and the industrial marketplace seem to imply a guarantee that they are being taken care of, that they will be taken care of on into the future. But 'it ain't necessarily so'. A lot of modern society has run on a qualified institutionalized faith. We trusted that our bank deposits were secure, we trusted that there wasn't any water in the fuel we purchased, we trusted that our jobs were somewhat secure, we trusted that the governent wouldn't allow some preventable outside threat to hurt us, we trusted that the tap water was clean, that the air was breathable, that the food was safe, that the pump would go on running forever. For some, to do otherwise was to go stark raving nuts with worry. But there were, and are, no guarantees. Today is a day when more of us are feeling this reality. Today is full of the doubts about tomorrow.

People out of work and keen, if not desperate, about surviving may constitute big opportunity for the country and the world. No one wishes for or welcomes their trepidation, yet I do see these folks and their need as an opening for constructive change. They want to work, so offer them jobs to rebuild our country's crumbling infrastructure AND offer them a chance to farm a piece of ground, offer them homestead opportunities. Such ideas might anger an out-of-work auto worker or stock broker but those stressed jobs may not be available for a very long time. In the meanwhile, good work rebuilding the country or farming will offer them a dignified alternative. Humanity needs fed. Corporate industrial agriculture tied to banking CANNOT do the job. Small independent farms scattered across the landscape, they alone can do it. We need millions of small farmers RIGHT NOW. I say we press the federal government to release large tracts of arable public land to a whole new homestead act.

> *It was summer time and the irrigation and pump repair guys were sure to be busy. It would cost me an arm and a leg to get anyone to come out here to service this unit. It's a big one, got to be 600 pounds all told. Luckily it's sitting above ground and in a solid little shed. That will give me a few options. What should I do? I could separate the pump from the motor to lessen the weight of each half but that would mean doing*

a blind surgery on the guts of the cast iron-enclosed centrifugal pump. I would need to trust that removing the nut on the end of the shaft to release the impeller would be sufficient to allow me to pry the two halves, engine and pump, apart. Not knowing what other pieces might be out of sight and potentially ruined, I hesitated. Then I remembered the adjustable pump-shaft collar that compresses the packing. I could get to that, so I removed it to check the status of the packing material. It seemed to be fine. So, that meant I was left with disconnecting the pump entirely, lifting it into my pickup truck and taking it 30 to 45 miles to a repair shop.

But first I needed to find a shop I could trust. So I called a few friends and found a good recommendation. I called that shop and the lady said that 'the repairmen are gone all day everyday doing installations and repairs.' If I wanted to bring it in she would see if they could do a rush repair.

"I can tell you one thing," she says, "it counts in your favor that you are not afraid to take it apart and bring it in yourself. These days we get so many people who don't want to get dirty and are afraid to try anything mechanical. The guys like someone who's not afraid to get greasy, oh, and I might add, also not afraid to admit when they don't know everything. That's what's moved into this area, a lot of people who claim to know everything and can't do a thing for themselves."

Today common sense, basic skills, and a self-sufficiency mindset are positioned to once again become each individual's most important tools. I am speaking of all aspects of daily life in this new confused century, but most particularly agriculture. So much of farming is done in tight places, trying to get something apart or back together again. Any chance at all of success translates to a measure of self-sufficiency which is self-fertilizing. What I mean by this is that each time you succeed with the mechanical, animal, crop, and/or procedural challenges of farming you feel thrice emboldened to deal with the next challenge. The success makes your self-confidence grow as if fertilized.

I went into the little pump shed with wrenches in hand and took everything free, especially careful to cut the power. I have a pair of heavy steel loading ramps for my equipment trailer. I took one of those and set it into the shed and on to the bed of the truck at a steep angle. Next, using a pry bar and three short lengths of pipe I raised corners of the heavy pump motor and slid the pipes in to make a roller bed for the motor bottom. Then I carefully rigged a short triangulated chain yoke on the pump and hooked it in to a come-along I had fastened at the back of the truck. Going slow and double checking all anchor points, I ratcheted that pump up into the truck bed. The whole thing took three hours, what with the extra cautions and running back and forth to the phone. I felt mighty good about getting it loaded but I had a passing worry about the time spent as I should have been working getting office work done or on training my two stud colts. I had to chuckle, these were minor worries compared to what was happening around the world, to the world, and in spite of the world.

I travel often and talk to a great many people. The conversations today, each today, start with shaking heads and worried glances as everyone acknowledges their worry over the economy. Some of us share talk about stocking up and fixing things so's we can take care of our families and friends when things go further down. A few express downright panic as they see they don't know what to do and are sure as heck not prepared to take care of themselves, let alone their families and friends. A few of us see these difficult times as opportunities, not to take advantage of the downtrodden but to provide for them in ways which help us as well.

Back since February and March, when the fuel prices grew springs, the middle eastern wars became institutionalized, the planet went into its shake and sweat, domestic inflation went through the roof, and the banking crisis announced it was here to stay, people began to mutter, chatter and worry. And as more and more factories and corporations laid off people the economy started to take on the causal stink of the greedy and incompetent past and present administrations.

The pump shop I took my unit to has been around a very long time and the distinctive galvanized building told you so. Driving in with the pump I noticed that the old trees had been cut down and the shop squeezed between two big new commercial developments almost complete but dangerously barren because all evidence of the contractors was gone. Another casualty of the banking crisis. I talked with the pump folks and explained my irrigation worry and the need for a quick repair. They were keen and promised results in two working days. Feeling like things were going to work out fine, I took a moment before loading up in the pickup truck and gazed around sighing, across the highway was a new Super Walmart, next door a four story resort motel half done and vacant. It was inevitable that this family pump business, supporting so many farms and homes, would inevitably be squeezed out. Another unfortunate change in the fabric of this community. I knew I wouldn't be bringing my irrigation pump to Walmart for repairs. These developments are not fed by need, they are spurred by speculation and they have already destroyed sustainable aspects of this community.

The word sustainability is used so wrecklessly and politically today that its meaning is perforated. Sad, because we need the word or its essence to hold water for us as we work to define and understand right livelihood and the human future on this planet. We must understand that true sustainability, that capacity for systems to regenerate and sustain themselves, is at war with the gods of commerce and the corporate ethic. And in true Machiavellian-style the enemy is hard at work to usurp the word "sustainability" as its own, redefined, retooled, and priced to sell.

The British essayist and mystery writer G.K. Chesterton warned sixty or seventy years ago that if we weren't careful, advertising would replace organized religion as the shaper of human society. Turns out he was right. We weren't careful and it did happen. Today, millions of people go to their grave believing that their individual life had been measured by their purchasing power. There are churches today who preach that God wants us to be commercially and financially successful. I disagree. I believe God wants us to be farmers, stewards of the land and of biological life, and happy campers because we feel good about our workaday world. I refuse to believe that God cares how many homes you own or what car you drive or the nature of your stock portfolio. But there are those who will argue with me in that direction. From atheist to agnostic to raging secular hedonist we argue with one another on these questions ... and they do not matter in the final analysis. Fact is, we can care about the world we live in or we can disregard her, it's our choice. Really our choice because those codes of conduct, moral institutions, and halls of learning which gave us the credos of caring, sharing and believing, those guides of old are gone. They've gone shopping - those churches, schools, synagogs, charitable institutions et.al., gone to the marketplace, left the stone tablets and ivy covered halls behind and stuffed their wallets with little magnetically charged plastic cards. Still, whether we care about the world we live in or not is something we, rich and/or poor, have power over. This is something we can see, touch, feel. And, if we allow it, it is an involvement which can reward our peace of mind and our self-sufficiency many times over.

> *I returned in two days and picked up the pump, repaired, sandblasted, painted and ready to reinstall, all for half what I had expected I would need to pay. And smiles and stories were thrown into the mix by the pump shop folks as I was asked where our ranch is and if I knew so and so and how the crops were faring. Felt good. As a friend of mine sometimes says at odd moments, "We're farming now." And he's right, this is part of the intangibles of farming. It's moments shared with folks of common values, all of us appreciating who we are and what we do. Part of that right livelihood business, we know at these softer moments that we are who we want and need to be.*
>
> *I drove home and readied my tools to reverse the process of loading the three phase pump and motor back onto the foundation. Rigging a chain into the rafters I hooked into the motor casing and half lifted / half drug the unit down the steel ramp. Using a heavy pry bar, I walked it into position and bolted it back into the piping. My last piece of work was to rewire the motor. I thought I was being careful to replace the wires as I had marked them. I buttoned everything up, straightened my old back and smiled at my handiwork. Now to open the water gate and turn the pump on. Sweet quiet running motor and pump but NO water came out.*
>
> *I phoned a friend and he confirmed my suspicion, I must have reversed the wires causing the pump to run in reverse direction. I took the covers off and redid the wiring. This time when I pressed the pump switch six hundred gallons a minute pulsed through the 8 inch pipe. "Now we're farming!"*
>
> *Sustainability? Self-sufficiency? You want to know what these things are all about? Follow a farmer or rancher around for a few days.*

There is no other segment of society which offers endless opportunity for regeneration and sustainability; farming is it. All other segments of society and commerce are ultimately governed by a limitation either of resources or of profitable application or of both. Humankind needs all the product of the farm each and every day. And all the product of the farm may, if we organize the farming accordingly, be part of the net gain to fertility, community and planet. We may grow beans, while improving the top soil, and extending the life of a seed variety. We may produce milk, while improving the top soil, improving a livestock strain, and adding to the beauty of a landscape. We may put up hay while training work animals and improving self-sufficiency.

If we make the claim that farming is the answer and that we need more small farmers and that there has never been a better time to be a farmer, it's appropriate to ask what sort of farming we are talking about. I am want to differentiate between industrial-scale agribusiness and the craft of human-scale farming. I believe the latter serves all of us best. And I am most certainly not speaking of a return to some antiquated, nostaligic form of farming. I am speaking of a new farming, not in the distant future, but one which is already scattered amongst us today. It's a new farming whose limits are defined by values. I like to think of it in terms of the individual unit, one person's farm.

One man's farm: from its perimeter to its center never further than one might walk with a load.
One woman's farm: never less than the circumference of the dream it represents.
One man's farm: no matter the scale so long as it befit an individual imagination and endeavor.
One woman's farm: a perfect interlocking fit to the next woman's farm, or the edge of nature.
One man's farm on balance, always on balance. Balance not as of style but as of manner. That farm must be the size which fits the true manner of the man. And the farm will be the prize hard-won of the deliberate maintenence of appropriate creativity.

Why one person's farm? Why not the farms of many? The farms of many should not be measured by their outer collective boundary but by the energy of the pattern of shared perimeters. Measuring the boundary of all, of the many, gives the lie to larger singularities. It's the difference between edge defining size versus the internal weave.

In classical music, symphonic orchestration requires, demands, the composer own a level of intimacy with each and every sound, instrument, human performance aspect and sound chamber. There is no other way to achieve the powerful fertile singularity of composition. The same is true of one person's farm; that farmer must own intimacy with all aspects and components of his farm if a singularly fertile and constantly regenerative whole be the goal. At some point increased size and quantity require abstraction and with this, intimacy is lost. Music is lost, fertility is forgotten. Regeneration and sustainability become conceptual impediments to profitability. But so what? Profitability is no measure of permanence or fertility or art or sustainability. And, as we see today, corporate profitability is off in some parallel universe with little or no use to those millions who, as we speak, are starving.

With a manner of persuasion born from understanding, the fertility of one small farm may be as regenerative and sustainable as the wave action of the sea. Do we measure profit from that wave action? Typically, profit is a subjective measure of income retained. The wave action of the sea is a phenomenon. The growth and produce of the soil is a phenomenon albeit one a thoughtful person might orchestrate.

There is so much we might say of profitability and all of it apology or rationale. There is so little we might say of the craft of farming and most of it promise and wealth. But this we can say, those particular demanding labors of one man's farm require of him that his mind be fully on his work. And so engaged, this man becomes his work and the work becomes the man. The farm's work takes on the manner of the man, be he artful and/or responsive - and the man likewise, be the work tedious and/or challenging. And these manners blend to a defining balance the repeating final ringing tone of which is accomplishment, tedium, labor, artistry, responsiveness, gathered gardening, stockmanship, challenges, all of the farm's fueled future towards a chance at the exhiliration of right livelihood. Farming for life. Turn the words around and inside out. A life for farming - or - farming in order to live - or - farming to create life - or - farming to sustain life - or - farming to support life. It works every way.

Farming for life now offers the grand answer to what is the collapse of the grand illusion.

thirteen

what I'm looking for is...

I called my buddy Ed Joseph on the phone and his seven year old daughter Natalie answered,

"Hello, who is this?"
"This is Lynn. Is your Daddy home?"
"Yes." Pause.

"Is he busy?" Silence and then a hesitant
"No."

"Can I talk with him?"
Long pause.
"It would be better if you called him tomorrow," she said.

"Honey. Would you tell him I called and ask him to call me back?" I said. It was followed by a prolonged silence.

"It would be better if you called him tomorrow. Who is this?"
"Its Lynn."

"Lynn?"
"Yes."

"Wes?"

"No, Lynn."

"Lynn?"
"Yes."

"Who's Wes?"
"Tell Daddy I called, okay?"

"Who is this...?"

This was a very bright little girl trying to control access to her father, but also trying to determine just how important the call might be. Found out later she never did let Ed know I called.

TYING KNOTS IN THE BINTURONG'S TAIL

And that conversation reminds me of several things that have been happening to all of us during these last couple of years; this disastrous economy, runamuck weather, bizarro politics, and nasty amped-up efforts to control food production.

For example; that exchange with little Natalie brings to mind the dialogue some of us have been trying to have with the USDA on NAIS (National Animal Identification Systems). There have been those that have felt the threat of Animal ID to be the proverbial "line drawn in the sand," that moment when a stand had to be taken. If we could just get their attention and stop them here...? But the USDA wasn't taking our calls, or they were choosing to confuse us with other factions, and they were definitely wanting to protect their "Daddies" who in this case are the corporate food giants.

And then came the "Food Safety" legislation. As if we needed to be reminded that there is a wider cultural war within U.S. agriculture.

Well now we're being told the USDA "got the message". They figured out that we've been calling for them, that we've been trying to tell them who we are and what we want. And they say they decided to drop NAIS. However some of us are certain the same fight is still there to be fought, only now under a different name and against a whole new set of tactics. "Where's the front now?" We are asked. About the time we completely figure out the answer to that question it will probably change. But we have our strong suspicions that the war is being staged to pit amateur farmers against professional farmers.

I once had a magical chance encounter with an Asian Bearcat, also known as a Binturong. So long as he was convinced I was his undemanding companion all was peaceful and exotic, but when I played with the thick hair of his tail I soon discovered a dangerous emotional landscape. Friends and fellows who are recent entries into the world of agribusiness and status-quo industrial farming are wondering at the ferosity they are met with by public and private protective farm institutions and agencies. Big agriculture is feeling threatened, they don't like us playing with their tail hairs.

CLEANING THE GORILLA'S MIRROR

For nigh on forty years I have wondered how it could be that the United States Department of Agriculture so easily and repeatedly forgets it serves us all - the U.S. Citizenry and, in many cases and ways, most particularly it serves farmers - the prime constituency. Yet even a casual inspection will show that a handful of corporations, academics and bureacrats get 99% of the attention and control 99% of the federal ag policy agenda. But something interesting is afoot. Now, within the less than hallowed halls of the USDA (wouldn't it make a grand parking garage for the national capital?) we are hearing mumblings. Seems the Ag. Depart is fixing to call us in for meetings and picnics, we individual farmers, and see if they can't win us back. They are finally figuring out the extent to which, with a wholesale loss of constituency, their very future is at stake. They don't get it, they can't see the larger pictures. And the world is changing too fast for their pleistocene ag economists to even calibrate.

WHAT FOOL WOULD ROCK THE BOAT IN A STORM?

Mindful of this hideous recession/depression we are now immersed in and the suffering it has caused so many fine folks, I was compelled - fool that I am - to point out worse days ahead. Speaking at the Placer Grown Conference in Lincoln, California, I made this statement;
"From climate anomalies to the collapse of industrial agriculture, from this 'great recession' to massive regional crop failures, from water shortages to weather-delayed plantings, from the increasing affluence of Asia to the disappearing wealth of the U.S. and more, all the factors are in place pointing to an

unprecedented shortage of food worldwide, and within 24 to 36 months."

I made the statement as part of a stage-set to hopefully drive home the conviction of most within the scientific and humanitarian communities that we may be faced with global hunger on a unparalleled scale. The evidence is mighty convincing. I don't want to believe it. But not to take heed would be incredibly foolish.

When I made the above statement in my keynote address there was an audible pause, almost a gasp, in the auditorium. It was as if everyone stopped breathing for a few seconds. And not out of surprise but perhaps because I had said out loud what many of them knew or suspected but didn't want to hear. "Maybe if we ignore it it won't come true." A few of them had perhaps even mistakenly thought I was too well-mannered to say such a thing.

On I went talking about challenges and opportunities, the need for gratitude and gumption; the rewards of grace and generosity. And I had clearly in mind the solid impression that had been made on me, moments before my talk, by the acceptance ceremony for the Placer Grown Farmer of the Year, Brian Kaminisky. Someone in that intro spoke of how each year Brian would have a customer appreciation day and up to 300 people would show up on the farm. I felt a seed of an idea planted in my brain; had no way of knowing it would germinate within a couple of moments.

Immediately after my keynote a woman stepped up and nervously asked "Can you tell me more about this world hunger business? My grown children think I'm nuts when I talk about it. I need to convince them, and I need to be able to give them ideas of what they might do to protect themselves."

I invited her to the question-and-answer workshop which followed and promised to answer her first. Then Brian Kaminsky stepped up to offer a sincere thanks for my talk and I said, "Congratulations on your award. I know you will appreciate when I say that everyone is a winner. In making this aknowledgement of you and your efforts, this community is also justifiably congratulating itself for its understanding of true wealth. Gratitude all around."

DUSTING OFF THE PEDDLAR'S BOAT-SHAPED WAGON

In that moment somethings coalesced in my tired brain. It was time for the question and answer workshop. I told the worried woman, after giving her and the others in attendance my chapter and verse facts to support the food shortage contention, I told her that the best thing she and her offspring could do to prepare was to develop strong personal and customer relations with at least 3 solid local farmers. Know, in a pinch, where you'd go for a dozen eggs, a gallon of milk, fruits, vegetables, meat. Make folks like Brian Kaminsky part of your family. I could feel others in the room nodding in agreement.

The next question, from a gentleman in the room, had to do with the many appropriate uses of the internet and searchable databases to bring farming information and farm community building into the new fast lane of society. I held my concerns in reserve while talking about our hesitant new efforts with social networking and website content. But in the format of this writing I do not feel I need to exercise the same caution.

I use the internet a great deal. While it is certainly true that it has replaced Encyclopedias, the Thomas Register of Manufacturers and the local librarian's card catalog, at the same time and with each passing day I find that the shear ready mass of cyber data has dramatically reduced the beauty and relevance of the material. Context used to be such a defining force in any discipline touching upon a reliance on craft and artistry. Farming, music, gardening, plant propagation, architecture, writing, and teaching (to name but a few) all were beholden to environmental, historical, peer-age, and community context. And those contexts, once welcome edges and forms, now with the internet have slipped around the ankles of our society and time. So little preparation, very little forethought, and all in the biggest hurry... but to what end?

Yes, there are many new ways to gather and send out information. And, maybe, with time these electronic databases will mature in their service to mankind, but I doubt that will happen without our insistence.

WE DON'T NEED NO MORE STINKING NON-PROFITS

Many thousands of miles were covered over a recent twelve month period of time to speak on behalf of a non-profit effort, and to listen to what people had to say, what they had to ask. And it was the questions that surprised us most, surprised and educated. But it took some doing to understand not just the questions people were asking but the new expectations stemming from new collective context and experience.

The collective context? Over these last twenty years we've all been hit repeatedly by the media blitzkrieg, a scattershot of commercialized political, charitable, and religious sales pitches swirled in with the beer, video game, pharmaceutical, showbiz and cell phone torrent. (We hardly had time to catch our breath before it all started to fall down around our ears when the banks left town with our money.) But all of that blather came at a time when many of us were learning how to use the internet to sell our old model trains, discover the truth about our misbehaving automobile, get cheaper stuff, and otherwise justify our growing scepticism. The key result of these last twenty years has been the evolution of a heat-tempered impatient populace anxious for short usable answers and poised to walk away the minute you bore them. So it was from this thick cold soup of discontent that we were served observations and questions.

Patterns started to appear, priorities seemed to appear and they weren't what we expected; "what I'm looking for is…"
1. (and by a long shot) where can we find …
(take your pick)
a. grass-fed meats?
b. raw milk / whole milk?
c. organic vegetables and fruits?
d. organic grains?
e. butter and cheeses?
f. organic wines and beer?
2. where can I find a job on a farm that will teach me what I need to know?
3. how do I afford a piece of farm land?

That's it, the top three, nothing else even came close.

And we thought we were 'speaking to the choir', that our audiences were people who would be more concerned with insurance, farmland preservation, marketing of their produce, etc. etc. Instead the majority were people who wanted entry and access. They wanted to know where stuff was - period. They didn't want another non-profit, they wanted answers - not information - answers. And right now, no beating around the bush.

"Give me the address where the grass-fed lamb is sold, now. I don't want a lot of talk, I don't want to join anything, I'm not interested in the politics, don't send me to the supermarket - nothing there is honest - and I don't care about the USDA certification crap because that's all it is. I don't care who the good guys and bad guys are - you know - cuz its obvious. I just want the real stuff right now. If you can't give it to me I'll find it elsewhere."

No, hello how are you's, no thank you's, no goodbye's, just 'put up or shut up, right now'.

Tough crowd. And about to get a whole lot tougher, especially if the indicators prove right and our food supply diminishes rapidly. There are some odd designer questions that will be asked soon; such as, will hungry people care if their food is organic, or heirloom, or GMO-free, or hormone free? Will parents anxious to feed their children care if the food is absolutely safe? Dumb questions? Think about it. I say they will care intensely because, along with the threat of food shortage (and some will argue it is a principle contributor to that shortage) will be the fact that industrially-produced food will be more and more susceptible to egregious behaviors in factories and farms. If you think the Toyota recalls are a big thing, you ain't seen nothing yet. Food poisonings are going up, no way of stopping it, unless honest and strong efforts are put in play to control food factories and factory farms.

Tough crowd, especially if their demands for access to land and farming skills go unmet. These folks are liable to take matters into their own hands and GO OUT AND PLANT, heaven forbid, LAND THAT DOESN'T BELONG TO THEM!?

Calm down, you say? But that's not me saying those things. I am the one who'se saying that these days (and ways) offer to us farmers and want-to-be farmers incredible opportunity, opportunity not to be wasted, opportunity to be understood and monitored for rapid shifts so that we can be there at the head of the line with bushels of asparagus, homemade cheese, a book on raising chickens, a ventilated bottle of ladybugs, and a smiling nod.

By the droves, we have people who need what we have. If we be farmers they most definitely need the food we produce. If we be folks who want to farm THEY definitely need us to be farming. There's a formula in there someplace! We need to see the magical mathematics of true supply and demand - not this gummied-up big-dogs-only corporate fascism we have today.

And you know what keeps today's system afloat? Efficiency? Nope. Good management? Definitely not. What keeps it afloat is our long held insistence that each of us go it alone, and our new found impatience with anything but instant gratification. Civility has been replaced by sequestry, community has been replaced with our new empty sugar-free greed. So we set ourselves apart with hands-held-out-impatiently and corporate fascism continues its massive bloat.

We could correct the course of our collective ship of fate in an instant if we could but all work together on common cause. Those obscene heads of the corporate food system are betting we won't. They are betting they understand that our baser instincts define us. And, you're right, we don't have to talk about it. We just need to do it.

A STOLEN ROAN GATHERS NO MOOSE

Everywhere I go I meet people who can't find what they are looking for. And I turn around and standing right behind me are people who have what the others are looking for but cannot find them what needs it. Gives people like me a sense of purpose. To connect those folks, you folks. Maybe that's exactly what we need to do, find the best and most popular ways to let everyone in your community know for certain that you have milk and eggs to sell. And let everyone in your community know you want a piece of land to farm and someone to show you how. I am convinced that we have all of the components, all of the answers. The opportunity is defined by the challenge to connect the dots, to find "convenient" and "obvious" ways for people to find what they are looking for. The only thing in short supply is the will to work together.

There is a favorite shaggy dog story that culminates with the punch line "A stolen Roan gathers no Moose." It has to do with an old Canadian rancher who rides out on a good roan horse and gathers Moose to sell. His envious neighbor figures the secret is the roan horse, so he steals him to gather moose but the horse won't work for him - ergo, the punch line. May seem a little convoluted but that's how I see the internet as an opportunity to connect us and our needs. The internet is like that roan horse, not much good without the proper cowboy (or is it mooseboy?) If the business community thinks it can connect us all simply by plugging in via some new or established web data base or program it just might be missing the boat. Knowing what rennet is, understanding microbial bacteria's function in soil health, valuing the fertile life of certain livestock, these sorts of things must come from the proper experience driven context. When it comes to farming they need to be in the background of any concerted, and hopefully sophisticated, effort to connect supplies with needs.

PARKING THE WATERMELON WAGON ON THE STREET CORNER

The regional history book tells of how the old Greek homesteader who first settled our ranch grew watermelons year after year, no irrigation, on a sub-irrigated hillside. He loaded those melons into a farm wagon back in the early twentieth century and drove several times, each consecutive year, 16 miles over land to park on the street corner in the tiny town of Sisters hawking the fruit. It was all so remarkable that it made it into the historical backdrop of the area.

Everyone knew to expect Sam Pappas and his melons right about that time of the year. That's what we're returning to. That's what we must return to. A community identifying itself with when the crop is available and where to go and get it.

Sequestered and insulated from true community, today we need a little help re-establishing those connections. We've said it before, and it cannot be said often enough; The World needs feeding and small farms are the way to get it done. We can do it. We can head off that worldwide food shortage, one local community at a time - all of a time, all right now.

Time to hunker down and make it all work.

fourteen

to market
to market
to buy a fat pig

Back in 2008, a farming friend declared that in the Seattle area the concept of local farmers serving local consumers seemed to have reached a saturation point. She observed that sales were the indication and to illustrate spoke of her own case; with farmer's market sales down two consecutive years - 20% down two years ago and another 20% dip in 2010. She suggested that while they might be needed elsewhere, at least in Skagit County Washington they did NOT need more farmers. In her view more farmers would just mean less income for each. Back then I respectively disagreed, and the same goes for today.

Several things were and are at play here, and together they spell confusion.

First; the great recession - sales continue down most EVERYWHERE. (And the big dogs of commerce, watch their sorry and insidious scurry for advantage.)

Second; three years ago URGENT and FASHIONABLE demand for fresh clean local foods coupled with lots of cash sales resulted in a most unfortunate attitude of entitlement on the part of some local farmers. Entitlement as in "I'm not going to chase sales - either they want this produce or they don't. They know where to find me. Don't these people know that they have to step up and buy this stuff if they want us to continue farming?" This is a dangerous attitude in the best of times. In this economic climate it borders on suicide. Good, creative, personable, aggressive sales efforts (including appropriate pricing) ARE resulting in respectable sales. Adjustments must be made constantly. Extra effort is making a difference, especially with today's tight purse strings. Standing and glaring at the passing customers doesn't make the grade.

Third; consumers in developed(?) countries will always search for convenience. And large corporate retail chains compete very hard to provide that convenience. If the consumer can get "fresh? local? organic?" produce from Safeway, while they are picking up the processed foods they can't live without, why would they go the extra miles to a farmer's market? It is another sign of success muddying the waters.

Fourth: The real dillution in the fresh local organic markets comes back door from the ill-advised and mis-applied USDA organic certification which has lowered standards and allowed large scale industrial production to flood the conventional markets with "fresh", "local" and organic" stamped goods which are highly suspect. Dangerous circumstance. And there is very little public notice of this. It is another sign of success muddying the waters.

Fifth: Within so-called alternative agriculture circles there are turf wars abrew with clubs, associations, coops, and marketing organizations all muscling-up to control the number of selling farmers and the amount of produce in an effort to protect "market share". It is another sign of success muddying the waters. Counterproductive to say the least.

The simplest applications of the so-called law of supply and demand suggest that there is some illusory optimal 'balance'. And that this balance wants to fall short. In other words if you have customers for 46 baskets of asparagus you need to have only 40 or 41 available to sell. This keeps demand sharp and price strong. If you have 50 baskets to sell and customers for only 46 the unsold will bring down the return to the farmer. But all of this works within some abstract finite notions of demand and customer count. The so-called law of supply and demand in our world of food production is an insidious apology, an excuse, a rationale for those who have already convinced themselves it can't work. It's destructive nonsense. With hundreds of millions of folks starving here and abroad, we have no moral right to speak of controlling production to keep "easy" and lucrative sales coming.

Even with only 50 customers in attendance, the farmer who has 75 baskets of incredible, tender, tasty and artful asparagus - and offers them each and all with a smile, a free soup recipe and a sprig of cilantro - will not only sell all of the produce but likely bring new customers to the market AND create a vitality that will entice new farmers. New farmers with increased urgency and goodwill bring more customers and some of those buyers will actually be farmers themselves.

The farmer's market with a couple of farmers sprinkled in amidst a handful of crafters does not attract the number of customers that a market of many farmers does. And the savy consumer, the true epicurean looking for that ultimate produce at a killer of a price, will recognize that the wide variety of many sellers suggests great deals can be made. It would be counterproductive if stalwart market farmers were to work to limit their number because they wanted to hang on to "market share".

If you happen upon a vendor with a card table displaying 4 sacks of homegrown oranges at a farmer's market where 6 other vendors are showing off some soaps, knit hats, and assorted salad greens, would you be likely to make a concerted effort to return to this small market? If instead you happen upon a lively, active populous market where you have to wait your turn to even get close enough to see the mountain of oranges cleverly displayed in the back of an old truck adjacent to a booth presenting a wide variety of mushrooms alongside homemade pastas and that smack up against an adventurous and expansive display of stone fruits rung round by jars of preserves reflecting the silvered lights of adjacent fresh iced fish which are in turn absorbing the soft greens of mountains of brocolli, spinach, lettuce and cabbage all softening the sounds of the guitar music flowing from behind the display of herbs and nuts, would you be likely to return? Even if, or especially if, your cash reserves were low? There is a reason why, through the ages, the vibrant street markets of the world have always attracted folks poor and rich. One reason is that because here we can see and feel the truest pulse of an ever changing supply and demand, a pulse which informs and decorates our living cultures.

Farming is the singlest greatest invention of mankind. After farming and certainly because of it, humanity's next greatest invention was the marketplace.

I've got two extra sacks of potatoes I don't need - I have a sense of what they are worth to me - maybe some measure of what it cost to grow them. I have a hankering for lamb and eggs but am not set up to produce them on my little market garden plot. To buy that stuff would set me back a bag of coin I don't have. So I set out with my potatoes to join lots of other farmers at a predetermined location where, each week, people come to swap goods. Once there I find myself tempted to swap for a fat little pig but, lucky for me, I don't have what the pig farmer needs. Then I notice a scuffle as three women fight over a small display pile of San Marino tomatoes to the exclamations of the farmer/salesman who publically laments not having grown more of this savory fruit. Short time later I see that same tomato farmer leading that fat little pig home. A light bulb goes off and right there and then I decide to grow San Marinos next season. The marketplace made it happen. Independent farmers gathered together to swap and sell holding a living breathing heart pumping economy in their hands. A real and familiar economy. Not that tall, dead, contrived thing we know as the "general economy".

It is too easy to blame it all on the general economy, but certainly there is the inescapable truth of this financial earthquake we all feel. We do ourselves a great disservice if we continue to think of this time, this great recession, as some natural phenomenon, some purely unavoidable correction in the general economic system. Yes, the balloon got too big and burst but that too was not inevitable. Though the opaque blankets and smoke screens may lead us otherwheres, I personally believe that what we are suffering through has been laid on us by a political and banking system gone berserk with greed, corruption and dishonesty. (And this has been allowed and even encouraged by the fourth estate, our press. Journalism no longer is journalism, its former practitioners have taken cheap seats on the bus to pleasure island.)

So what has this to do with us, a farflung community of small farmers and folks who care intensely about small farms? Maybe just about everything.

The world needs millions of new independent small farms and it needs them NOW. What's more, regardless of whether it is Skagit County or a rural county in Bangladesh, Los Angeles or remote Iowa, more farmers - more farms - and more produce will energize local markets everywhere. The industrial model of agriculture is failing in very big and nasty ways. Nasty weather, commercial fertilizer shortages, protection rackets, genetic mutatologies, wholesale industrial breakdowns, and many other elements are going into a mix that is aggravating the rapidly approaching worldwide food shortage.

Once I was in a large supermarket chain store and saw printed banners in the produce section warning consumers of coming shortages in certain produce brought on in part by weather problems. Those banners made it very clear that there would be less lettuce and the price would be substantially higher! I suggest we will see more such banners cropping (pun intended) up everywhere.

I sit at my kitchen table, early in the morning, and look out the little windows at a completely altered landscape. Last night it snowed eleven inches, and today the familiar has been altered. Everything is weighed down, lidded by the quiet white cloak, soft and silent. Animals wait to see if this is temporary, to see if they need to remind themselves of old useful postures and routines. Wild rabbits wait in familiar little hidey holes as do the coyotes. Winter birds are backed under their covers glad for the wind to be gone, not yet worried about food. It is a pleasant time and lovely too. Twenty degrees before the sun is up, but who knows if today it will burn through the low fog thickened snow clouds? Who knows if it will continue to snow or warm up? Who knows if it will turn this beautiful ground cover to a raging destructive flood? Or allow it to remain for months as the first blanket of many? What we do know is that overnight what we took for granted has changed. We took for granted being able to walk leisurely, if sometimes cold, to the barn to initiate chores. We took for granted that we'd be able to find that shovel, that hose, that chain, easily. And now it is buried and hidden. This moment just after the big snowfall can be seen as an analogy for this economy and the difficult and different moments we find ourselves locked in. We wait to know how to behave. If we are farmers we may even be thinking about ways to tighten up our sales. How do we sell more of what we produce? Do we need to dissuade others from joining our ranks? No, not if our markets and the vitality of our cultures matter to us.

Why do farm markets matter? How is it that they insist themselves, haphazard - vaudevillian - corny - burlesquelike upon us? What do we risk by continuing to discount one of the oldest and most organic of social service partnerships? These questions go to the core of what I suspect is a deeply ingrained flaw in the human psyche. As we progress, or suspect we progress, up that transparent ladder towards questionable social supremacy we are quick to shed those skins that would paint us as ordinary, slow, common, dirty, and trapped. "Farming" is one of those things we seem too quick to shed. "Market bazaar" is another. So, it's not just about our sales, it's also about how we want to see ourselves in society. "Farming" that original nobility and "market bazaar" the time-tested entry towards security, gave us this day and our daily bread. We need to constantly remind ourselves of this. We need to see ourselves as worthy. We need to always be ready to invite others into our ranks.

One thousand years ago a farmer's market was established for the city of Paris. Today it is known as the Rungis International Food Market. Huge, bustling, modernized and archaic (when measured against the corporate world), it spreads out today to exceed the landmass of the principality Monaco! Up until 1969 it operated out of the heart of the great city of Paris and was fondly known as Les Halles. Today, out of the necessity of scale, it is 4 miles outside of the city near the village of Rungis and includes its own beltway, railway, banks, hotels, car rental stations and truck repair shop. It is its own city - a city that lives at night. Even extending to its more recent digs, Les Halles / Rungis has for ten centuries been the belly of France - the driving force behind one of the most dramatically epicurean societies in history.

Rungis is not a thing in and of itself, it is a thing in the aggregate - and that is a critically important distinction. Here is an example of how many independent produce and market ventures have come together with an economic vitality that has arguably shaped an entire nation to beautiful advantage.

Rungis is a fresh market, first and foremost. At 2 am buyers arrive in droves to select - from acres of giant, connecting halls - meats and cheeses, fruits, vegetables and flowers. Rungis feeds 11 million people in the Paris region every day, as well as supplying markets and restaurants around the world. Eleven million! Can you imagine having a discussion of how to limit the number of farmers delivering produce to Rungis so as to hold prices steady? Unthinkable. Have prices swayed? Of course they have. Does everyone benefit from the market being truly free and open? You bet they do.

Day in day out for over 1,000 years the farmer's market of Les Halles / Rungis has had as much a hand in the evolution and development of French society and the city of Paris as any other cultural or social aspect. It is sobering to think what might have been had these markets been asked to "temper" themselves with a view towards protecting that original band of farmers, naturally concerned about their income levels. The French might say "let them fight for their corners for with that we will add more corners, many more corners. And we will grow our city, grow our region, grow our culture!"

Rungis market is big, very big. And some might observe, too big. But it is not the only farm produce market bazaar. Others, much smaller serve neighborhoods, and these in turn feed off the success of all the others big and small. Overlapping concentric circles, small markets, medium sized markets and big markets. All of it is about economic vitality and the intrinsic independence of millions of healthy small farm ventures.

Instead of fewer farmers at farmer's markets, we need MORE hustle and bustle at farmer's markets - we need MORE farmers MORE variety MORE opportunity MORE thrivance MORE corners, many more corners. Perhaps then we can look forward, over time, to having recreated a genuine culture of depth and happenstance for ourselves and for our futures.

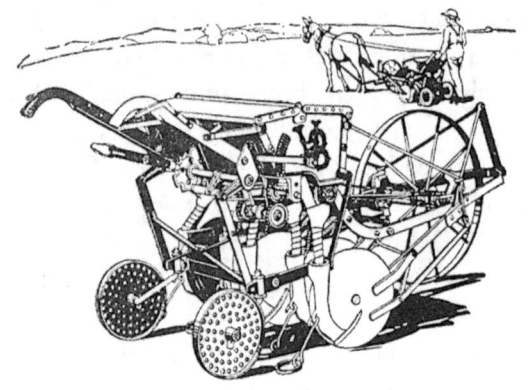

fifteen

skunks, snails and...

In one hand five wrenches, 3/8", 1/2", 9/16", 5/8" and crescent, in back pocket a rolled and dirty implement manual, I walk that quarter mile back to the hayfield where I left the busted mower. I know what I need to do now, so I enjoy the luxury of thinking wide. It's quiet here, quiet and lovely. No neighbors, no traffic, no sign of any insistent commerce, just a skittering coyote pup racing away from my approach. So I think of how fortunate we are and, try though I may, I cannot keep out of my old brain snatches of the terrible stories of starving Somalians, riots in England, wars and unrest in the mideast, tent cities in Haiti and across the U.S. as homeless jobless thousands wonder after a world gone stark raving nuts. But before the staggering list can be completed, I am saved because the distance to the mower has been covered and I am back to the work at hand.

Grease blotches scattered in odd places over torn clothing, dust hiding in the folds of the neck, ears, small of the back - jagged fingernails - a steadying fatigue - I hold a clear focus on the farming job at hand. A long lifetime of working at these jobs and sharing stories of other people at these jobs, I know that every moment begs to be watched and understood lest that one accident, that one quirk of misaligned movement results in breakdown of implement, animal and/or farmer. It is a vigilance that may hold us fast and firm and competent. Without it we are a bit of chaff in the wind of the work. We may survive today, tomorrow or the next days but we are, without that knowledgeable vigilance, just stealing a ride. Nothing earned, even less appreciated. Ironic that for me, this chosen, calm, steadfast vigilance gives me an arm's width of purest freedom. I don't want to be anywhere else. This is my moment.

And I know there is wider value in all of that, I know that efforts to do my best with this alloted slice of farming I hold adds in a very small but significant way to a better world. But, I have always felt drawn to those terrible pictures of inequity and ruin with a compunction-driven urge to do something more. Don't know, even at this late stage, exactly what that means. But I see things, and feel them too.

I feel that our communities and neighborhoods are falling apart. I see that there is less and less evidence of any collective knowledgeable vigilance in our communities - and that absence is beginning to be felt in neighborhoods as well. There is that piece of a line in the Declaration of Independence which has always stopped me to think... 'we hold these truths...". We did, sometimes well, sometimes not so well, but we did. We held those truths... Seems now we don't so much. We have come to take for granted the most basic of truths, the obligations of family, the possible sanctity of work, the human as a humble piece of a biological universe, the realigning and reaffirming power of beauty, the deep comfort of great friendship, the absolute primacy of life, and the list goes on. The money-changers, the creeps of commerce, the death-dealing ethos of the board-room, the criminality of store-bought science, the slithering subjective moral insolvency of a plastic humanistic psychology which puts the individual ahead of everything and ahead of nothing, the economic crutch of war and the emotional poison of digitized social networking, all of these things have driven us from our truer selves.

Ah, but then there is farming, that handmade farming which holds us in the wider patterns. In the heat of summer, crops calling out, we know persistence. If we are able we wear patience as we carry our persistence. And with farming's required patience and persistence comes ample opportunity for clarity.

After all these years there's plenty of things I should'a known, certainly many little things about farming. But the older I get the more clear it is to me that I know relatively little. Take knotters and neighbors for example.

My buddy Ed recently told me something about baler knotters, I'm sure it must apply as well to binder knotters. And the information seemed a metaphor of neighborliness as compared to community.

This is the way it goes; knotters have adjustable twine discs which hold the string during the knotting process, allowing the proper tension for the tying. If the string is sisal or if it is synthetic poly twine different relative thicknesses apply. So it stands to reason that a setting for sisal might account for too much space between those discs to properly hold poly-twine.

I'm seeing those twine discs as an adjustable filter of sorts. We can tighten or loosen them to make a certain dimension of string work and resulting in other string being held off. Now I see that as an analogy for neighborhoods. Some, by their design and 'adjustment', favor certain income and social brackets - while others are set to prevent certain people from any comfort in passing through. I feel held off from wealthy exclusive rural communities, I just don't feel welcome, it's that exclusivity factor, the 'twine disc is set too tight'. Whereas a rural landscape peppered by thriving small farms and businesses welcomes me right in. (Twine disc set loose.) That's just me talking. Cranky ol' me.

You might observe that I should be speaking, perhaps, of communities rather than neighbors and here is where I would beg to differ. We, the wider western world, tossed the words around helter skelter these last decades and permitted ad agencies to make of them what they will but I wish to offer that 'community' by definition speaks to a group of like-minded, like-employed, or membership-bound people such as Carpenters, Orthodontists, Baptists or Shriners. Commonality is the guiding principle. Whereas 'neighborhood' speaks to proximity, shared environs, locality. Neighborhoods may contain a strong or absolute contingent of community, as in this is a black neighborhood or a Jewish neighborhood. But most neighborhoods are a mix of races, religions, sports preferences, fraternal memberships, income levels, etc.

Why these word distinctions might matter here and now? Some of our communities are solidifying, calcifying, thickening, getting stiff - especially as regards politics and cultural issues while many of our neighborhoods are becoming dark and forboding. Members of some church communities are becoming completely intolerant of other church communities. People in neighborhoods are feeling more and more threatened and/or are suspicious of others in their own locale. These are not good turns.

We are a community of small farmers, farflung yet sharing many values in common. We, each of us, are part of local neighborhoods, which may or may not serve us as a market base for what our farms produce. It is ironic that while we may know other small farmers, living far from our locale, quite well - we have neighbors next door who we know little or nothing about.

I, for one, see the biological environs of our ranch as neighborhood. The eagles, the badgers, the mule deer, elk, cougar, coyotes mixed in with the bitterbrush, sage, pines, junipers, rocks, bitterroot, sky, dirt - every piece of it part of our neighborhood. And I have complete conviction that we need this fragile balance that is our local ecosystem - and that we in turn contribute to its balance and vitality.

But our case is somewhat unique in that we have no actual neighbors for five miles in any direction. We are 'by choice' isolated. But are we really? Recent efforts to create a wilderness area bordering our ranch on the east side and encompassing a stretch of the Deschutes River have played up the question in a most specific fashion. We have signed on to the proposal endorsing its objectives. Recently while visiting the local hardware store a young man approached me and said he could not understand how we would go against our "neighbors" wishes by our endorsement. The implication was that we were ostracizing ourselves by going against the wishes of the wider group that is seen as our neighborhood. Though five miles away as the crow flies, that community rests on a mesa the other side of a 700 foot deep river canyon. "You can't get there from here." I've lived on this ranch for more than 22 years and I have never been to that community. Our towns are Sisters to the south and Madras, across the lake, to the north both at nearly 20 miles away. We do business in those towns and we share many neighborhood concerns with them because our own business is there in Sisters and we do our annual auction in Madras. But the community to the east, the one bordering the wilderness proposal, we are completely separate from them and yet...

Many if not most folks have personal situations which incorporate jigsaw puzzle elements in how their chosen communities and their landed neighborhoods set together on a map. We think we have a say in what's included or not. Sometimes encroachment changes definitions.

We have used predator control dogs to hold at bay the wildlife that would redefine our farmstead, garden crops, and livestock (in other words our immediate personal neighborhood). These days our Great Pyrennes no longer has the full use of his back hips and our old Australian shepherd has vision and hearing constraints. We will no doubt be replacing them soon because we feel each day how the wildlife have moved in closer. Where we seldom ever saw wild rabbits near buildings and in garden areas they now come and go at will. With them has come the pack of coyotes a dozen or more strong. Last week I shot and killed one within twenty feet of the house in the middle of the morning. (I don't hunt them as a rule, I certainly don't mind if other people do. Coyotes do a valuable service controlling the population of sage rats, rock chucks and, to some small degree, badgers but when they decide to move in on us with impunity, measures have to be taken. Enter the human predation element.) There is a natural balance to our neighborhood. How we behave with and within all the natural elements affects that balance.

The wildlife habitat we call our neighborhood extends to an area of about a quarter of a million acres, running from the Deschutes River canyon on the westside all the way towards the neighborhood of Camp Sherman and then around the north to the banks of the Metolius River. The south side is marked by housing developments and Whychus creek (formerly Squaw Creek). This area was prized hunting and burial ground for the Pauite Nation before the 19th century "discoveries" of Captain John Fremont and Kit Carson. The migrating mule deer and elk use this area for winter habitat, moving in good climets to the High Cascades which border. As all-terrain vehicles and human population pressures increase, the migratory patterns fracture and constrict affecting birthing cycles, feed supplies, and predator concentrations. In other words the balance is altered. We believe, and have seen how, the existence of our "low impact" ranch has actually improved wildlife habitat offering central watering and shelter during critical times of the year. Without the ranch's working presence we are certain that this "neighborhood" would deteriorate quickly. But in order to have all of it work as well as it does we need to do our part to selectively and constructively "pressurize" our presence. What I mean by this is that when deer and elk come to our ponds for water and our field edges for grazing they NEED to feel cautious and ready to flee. Otherwise several hundred elk and several thousand mule deer would decimate the forage of this irrigated corner in no time at all. We don't want them hanging around, comfortable within a protected area. We want them coming and going with normal caution, concerned about us and coyotes and cougars and great white hunters. It's a tricky balance but one that has evolved over these short hundred years of shared dominion.

All of this talk about our wildlife habitat realities and thoughts are meant to suggest that neighborhoods are seldom simple. The best ones are a constant balancing of communities with sometimes conflicting values. We recognize that our wish to provide refuge for the wildlife seems in direct contradiction to our need to control their full access to our farm fields.

All human neighborhoods contain, to varying degrees, these elements of balanced conflicting values. Ethnic tensions around the world point to the challenges of maintaining some semblance of peace in neighborhoods made up of warring racial, religious, and political elements. Communities within these neighborhoods want to completely insulate themselves one from another but such efforts cause problems that go to the heart of continuing strife. What is needed is for communities inside of neighborhoods to work consciously to at least understand how they differ from each other, to understand how a measure of mutual respect can be earned from sharing the local news and calendars. Here's an example of what I mean:

Paul Hunter, and I visited Hillcrest Orchard in Penryn, California. Steve Pilz, third generation on the farm, took us on a walking tour sharing his passion for the diversity and fragility of this lovely sheltered world. The farm sits atop a hill and cascades down its sides to a ribbon of encircling bottom land. On the crest of the hill is a naturally-filled reservoir supplying gravity-flow water to the citrus trees on the grade and finally to the market gardens down below. The waters are directed by buried tiles.

As we walked, Steve pointed down and said "watch out, don't trip on that broken tile. I leave that opening there because it provides a home for our friends the skunks."

"How", I ask," is it that skunks are your friends?"

Steve answers, "They are a beneficial member of the neighborhood of this farm, at night they come out and eat the snails. If they didn't I'd have to find some way to control those snails because they could do a lot of damage."

Our tour continues on down the hill as Steve talks about the intracacies of the biology of an organic orchard, telling us of how it is that the best fruit always comes from the oldest trees. In his orchard the maturity of the tree is prized for the quality of each orange even though it may not produce as many pounds of fruit per year.

Coming down the hill it is hard not to notice that this lovely old farm in completely encircled by relatively new suburban tract homes, so I ask the direct question; "Steve talk to me about your neighbors there. How do they feel about your farming right across the fence?"

"Most of them identify with the farm positively. Over the years their kids have been coming here for summer jobs. But it isn't always roses. You see that big house there, with the swimming pool? They don't particularly like us and I guess I can understand why. You see, in the early summer we have large loads of livestock manure delivered to use for compost and fertilizer. It was on a windy day when they had a wedding for their daughter in their yard and we had a big load of fresh manure dumped up here. The strong odor gave an unfortunate flavor to that wedding."

"Oh my," I said, "that could not have been good."

Steve smiled a sad smile and said, "They should have invited us to the wedding."

"I'm sorry," I said, "what do you mean exactly?"

"I would never have had that load delivered on that day had I known they were having a wedding in their yard. If they had invited us to the wedding we would have known."

There it was in a nutshell, the whole story of what and why neighborhoods work and don't. How can we hold what we don't know? Ironic that a chosen, calm, steadfast vigilance and concern can give us a neighborhood's width of purest freedom.

All of those thoughts coming to me in the farming; back and forth, across the field, slow monotonous back and forth until your old body says "I'm seizing up here - do something different or I'll never let you bend freely again." So you stop, get off, stretch and mutter to yourself. It's easy to be cranky and ungrateful. I am not interested in what's easy unless there is some fertile reward of lasting consequence. So, whether I'm old or not, I'm making big efforts to be positive and grateful. Because it sets the tempo for everything else. And truth is I am a most fortunate man.

Speaking of which, when friends made offhand remarks about so-called obvious signs of age, I reminded myself that I ain't old until I can no longer put my experience to good use. And yes, for some of us that happens overnight. For others there may be that ober-season of "muleing" your way through with nothing to guide you but your glued-on inertia.

To paraphrase my friend, the poet Paul Hunter, you should love the work that loves you back.

sixteen

fierce plowman

He held close, every day, the layers of his farm - the livestock, each species; the fields at their readiness or usefulness or at the fallow; the ripenings, the remainders, the margins, the rottings, the seeds, the pollen races, the droppings, the absorbent chaff, the everything of his, this farm world. Close as it all was to him it required and earned his attention. He could tell you what piece of that field had a shallower top soil, he could tell you the history of the grandmother of that Guernsey heifer and how it might influence the coming partuition, he could predict the bloom of different crops and talk of how the bees affected it all passing one to the other, he did speak of this strain of legume seed he had carefully gathered and replanted for a quarter of a century, and he could wax poetic about plowing. He loved to plow, loved the slicing of the earth, the flip, the crumbling curving wave, the evidence it allowed him. He never tired of 'working' his soil and having it work for him.

Great Uncle Ephraim farmed his whole life in Minnesota. His time spanned nine-plus decades from the post-civil war years forward. He was successful and solid. He believed to his core that he knew why he was successful, it was because he was a good farmer who trusted the evidence of his years and fields and cows. When America spawned its golden era of farming, from 1900 to 1920, Ephraim was there to absorb and apply. Most of his latter years were spent alone with his fields and his Guernseys. Those pre-chemical-warfare years of farming were rich in the profitable theories and practises of a many-layered and multi-tiered agriculture. Crop rotations, rotational grazing, and an applied respect for the finer moments of seasonal bio-rhythms made of his place an ever changing jewel of diversity. His was a complex approach, lacing different aspects together - the livestock were allowed and encouraged to compliment crops, cropping and soil management while the harvest of feeds always took into consideration the other components be they birthing, breeding, weather, or overall timing. For the intricate overlapping crop rotation cycles he employed, cycles that could run to six years, he designed his field sizes to advantage thinking in terms of 'lands' rather than fields and keeping those 'lands' at 4 to 10 acres maximum. Of his quarter section thirty acres were in woods and farmstead, the remaining were split in changing mosaic between pasture and crop land. He enjoyed giving pieces of his land three to four year holidays as pasture as much as he enjoyed plowing those up to bring them back into the cropping rotation. Great Uncle Ephraim loved to plow. In fact he would argue fiercely that what caused farmers to fail was lack of regard for the plow and plowing. In his last years he got wind of arguments against plowing, arguments which pointed to the moldboard as the thing which caused the great dustbowl. Those arguments angered and confused him, He didn't understand any of it and was quick to say "I don't know what I don't know, but here farming is working the land and working with the land". For him, if you were to farm in the hill country of Minnesota you had better learn to love the plow. Great Uncle Ephraim was a fierce plowman.

The Poisons Take it All

Jumping back a ways, with a longer view, we can speak now of how it was that the great war efforts and the fragile economy saw the inevitable spread of heavy chemistry across the agricultural landscape. When the two world wars wound down there had to be a place to apply the mechanization and chemistry no longer required in European trenches. So it was force fed and dumped on our advanced and once elegant farming systems. We've seen the results and they have often been terrible. Chemical fertilizers, herbicides, defoliants, fungicides, insecticides, and sterilization elements all killing and misshaping our farming. The growing of food and fiber went from art and craft (as in Uncle Ephraim's case) to industrial process and mining. The result has been a deteriorization of the environment, a diminishment of genetic diversity, a depopulation of the countryside and a reduction in our farm productivity. For most of these last forty years our sorry-butt political and academic leaders have argued that what we have is the best system of food production and that what we left behind was "drudgery, superstition, and poor yields".

What our industrial system left behind was my Great Uncle Ephraim, and millions like him, and he never saw his labor as drudgery, he never felt his beliefs to be superstition bound, and he knew his yields were outstanding. He had secrets to share, he had grounded fears to pass on, and he wanted to give to young people his love of the cows and of plowing. But that was not to happen. Not directly.

Forward to The Beginnings

Now today, out of and in spite of the wasteland that is agribusiness, we see growing evidence, even an avalanche of hopeful examples all pointing to a return to farming as art and craft.

I've seen the evidence, I know what it looks like, smells like, hums like. I'm speaking of the very best that farming can be. I am speaking of the trail and picture of consummate regard for the four dimensional musical composition that a handmade farming might be. I've seen it, many times in my lifetime. But recently I saw it nearby. The best farmer I know is Brian MacNaughton. He has worked for us for several years, helping at the ranch all the while doing his own postage-stamp-size market garden farm huge in its production and fertility. I bring up Brian's example because he is proof for me that the old ways, Ephraim's ways, have become new again.

And lest you think I am pointing to Uncle Ephraim's as the old way please allow me to point out that his ways were just representative of ONE culmination of an attitude and approach towards farming that is thousands of years in the making.

The Chinese author Chen Pu (also known as Chen Fu) wrote in 1149 "Nongshu" or "On Farming". What follows is an excerpt.

Plowing

Early and late plowing both have their advantages. For the early rice crops, as soon as the reaping is completed, immediately plow the fields and expose the stalks to glaring sunlight. Then add manure and bury the stalks to nourish the soil. Next, plant beans, wheat and vegetables to ripen and fertilize the soil so as to minimize the next year's labor. In addition, when the harvest is good these extra crops can add to the yearly income. For late crops, however, do not plow until spring. Because the rice stalks are soft but tough, it is necessary to wait until they have fully decayed to plow satisfactorily.

In the mountains, plateaus and wet areas, it is usually cold. The fields here should be deeply plowed and soaked with water released from reservoirs. Throughout the winter, the water will be absorbed, and the snow and frost will freeze the soil so that it will become brittle and crumbly. At the beginning of spring, spread the fields with decayed weeds and leaves and then burn them, so that the soil will become warm enough for the seeds to sprout. In this way, cold as the freezing springs may be, they cannot harm the crop. If you fail to treat the soil this way, then the arteries of the fields, being soaked constantly by freezing rains, will be cold, and the crop will be poor.

When it is time to sow the seed, sprinkle lime in the wet soil to root out harmful insect larvae.

Chen Pu lived in the midst of the Song Dynasty, a period of tremendous agricultural productivity. This period benefited from the refinement of double and triple cropping in irrigated fields made possible by new farming techniques aided by the spread of information. Chen Pu published handbooks on farming which were circulated across the country. It is said that the richness of the farming from this period resulted in dramatic growth and stability for China.

Today China is as much at risk as the U.S. of losing its productivity, heritage and biological diversity as it grants to global corporations the right to poison in the name of agri-business.

We still have access to much of the information that supported our best farming though we may have lost the direct living connection and handoffs from people like Uncle Ephraim. But do we have the will, as a people, to find our way back? I believe it may come down to what we collectively believe to be 'truth'.

Social Truths?

In our society, this time argues with us - each of us - that 'social truth' is trapped within a moveable constantly shifting and overlapping grid. It's almost as though 'social truth' has become a circumstantial oxymoron, that in this day and age there is nothing completely true or absolute about our society. Aren't we too various to be, all of us, of or about or dedicated to anything even the higher human pursuits? Can it be said of our society that as a whole it believes in the sanctity of life? Can it be said that our society absolutely values the natural world and bio-diversity? Can it be said that our society is on the side of spirituality? Questions of religious and political polarization as well as techno-artisinal spirit-wrestling are only pieces of a wider confusion that threatens to make of homogeneity a curious relic. We are, without always realizing it, allowing ourselves to be herded towards 'concentration' camps delineated by our chosen 'persuasions'. I am a painter, writer and farmer. By those choices I am being herded towards a prejudged social encampment of people who are believed to be 'concentrating' on 'liberal' and self-gratifying endeavors and beliefs. I am seen by many as some weird kind of aging hippy with no regard for the politics of others. That, in spite of the evidence that I've spent my life, talents and interests working in the opposite direction. That is only important in this writing as I make a case for all of us to allow the best evidence to affect our next set of choices.

Farming is at a cross-roads right now. Industrial agribusiness is a miserable failure which struggles to compound the damage it has done by desperately 'doubling down' its wager in the arenas of scientific mutation and chemical warfare. Millions of people worldwide want to farm and suspect, against all corporate propaganda to the reverse, that given half a chance they could make a go of it. So they wiggle around in corners such as this looking for answers, clues, road maps and evidence. When they find the stories of Chen Pu and Uncle Ephraim and Brian MacNaughton you can see the electricity. But they still fight that new bugaboo I call 'the acids of social truth'.

All through my adult life, because of my interests in things agricultural, artistic, and environmental I have found myself participating in committee discussions at seminal moments when well-meaning public-interest groups felt they had discovered the single-defining-argument, that piece of the puzzle which when applied would 'force' everyone else to accept the absolute inevitability of their prescribed agenda. And frequently it was apparent at that time that a risk existed that to push the argument to the absolute of that decision would change the rules of the game for a dangerously long time.

I was a member of two different organizations, one a non-governmental body and another a government task force, both of which were pledged to find a workable solution to the destruction of clear-cutting Pacific Northwest forests. We had mountains of arguments and examples of the incredible potential for sustainable forestry practises, with a 'farming' approach to the woods rather than a 'mining' approach. That was the direction we were working in when some of the people in our midst had what they saw as an *Aha* moment that would 'win' the game for all time. Few of us understood that in that moment we traded our work towards elegant systemic solutions for a court-ordered edict disenfranchising not only people but biological diversity as well.

The lawyers and tacticians in our groups pointed to the Federal Endangered Species Act and said "if we can find just one species endangered by clear-cutting we will win this in the courts!" I and others said, "whoa folks, do we really want to change the rules of the game to that extent? What happens in the future IF your endangered specie should be shown to thrive elsewhere? Or if it should be found to be extinct in spite of this effort? What happens to the forests then? You are saying that 'so long as this is our identified concern you can no longer clear-cut these forests' - you are inviting that "when this is no longer a concern it will be 'a Bob's Your Uncle' moment, simple as that, cut as fast as we can every tree we can 'find'!"

Find? Finding? We speak of findings, in this context, as what the research would evidence. Semantically it is most telling that it would suggest that frequently we have 'found' something that was lost, perhaps in plain sight.

Yes, do what you must to save the Spotted Owl BUT do not make the future of our forests to hinge on that misplaced rusty pin. Instead SHOW and PROVE and EXEMPLIFY how sustainable forestry practises will grow more trees, provide ample lumber and jobs, protect the fragile forest ecosystem and our environment.

Within this little aside of my connection to the spotted owl narrative I see the parallels to agriculture in general. In spite of the overwhelming evidence of what a rich inherited farming craft can give and has given us we still allow the linear thinkers in our midst to apply the bigger hammer at the expense of the mandolins. But that is no longer necessary BECAUSE we have hundreds of thousands of new farmers worldwide who have taken their initial enchantments with farming beyond implication and well into application. We have that strong shot at exampling and showcasing the elegant systemic solutions that ARE a craft and human-based agriculture. But still we must beware the *Aha* moments in the hands of the grandchildren of those architects of industrial agriculture, those who, while they shop at Whole Foods hold to the belief that the future must be shaped by police and government edict.

- Those who for wildly various reasons, want to outlaw the consumption of this or that piece of the food pyramid,
- those who point to manures and say they harbor disease and must NEVER be used as fertilizer,
- those who would, out of a concern for some empathetic connection to the bovine, outlaw the human consumption of milk,
- those who would mandate the dehorning of livestock,
- those who would outlaw the use of equine in harness because it is cruel,
- those who demand that grains not be fed to livestock,
- those who are on a mission to criminalize choice in farming,
- those who would make it illegal for amateurs to farm,
- and on and on...

I say all of this and more represents a body of folk with entirely too much time on their hands. Focus is lost. That focus that would provide some distance and clarity. While we pick at each other in these ways, multi-national corporations and store-bought science continue to mutate life and sell poisons that destroy the biology of this planet. While we nit-pick and divert, war-waging governments continue to endorse the mining of the world. There are important things to outlaw, we don't have time or excuse to mess with our neighbors. If we insist on keeping things close-at-hand, then we should be spending more time on our own farming adventure. What happened to the maturity of our culture and society? Where did it go?

I'm reminded of that line from the Dylan song, "we were so much older then, we're younger than that now."

And on that note

It took 9 years, three separate controlled experiments, side-by-side, conducted by a wheelbarrow-load of academics from the USDA, Minnesota and Iowa to determine dramatically and conclusively that we CAN affordably feed the world, improve the environment, grow the top soil, pay the farmers for the work AND end the addiction to expensive and destructive chemicals. And that was the unintended consequence of this research. So much so that some of the architects, most notably the USDA, hope it quietly goes away.

On October 19 of 2012 a food writer for the New York Times, Mark Bittman, (someone who has demonstrated a limited understanding of the culture of agriculture and a general disdain for small farms) took credit for announcing to the world that a simple fix had been found for farming. An *Aha* moment? This is how he opens his article entitled "A Simple Fix for Farming"

> It's becoming clear that we can grow all the food we need, and profitably, with far fewer chemicals. And I'm not talking about imposing some utopian vision of small organic farms on the world. Conventional agriculture can shed much of its chemical use — if it wants to.
>
> This was hammered home once again in what may be the most important agricultural study this year, although it has been largely ignored by the media, two of the leading science journals and even one of the study's sponsors, the often hapless Department of Agriculture.

Bittman references what we are calling the *Marsden Project* entailing 9 years of research in crop rotation systems analyzing a comparison of the industrial model of corn/soybeans with three and four year rotations. The New York Times has seldom bothered itself with any deep tissue analysis of our agriculture because it doesn't sell perfume ads. But it is more than notable that this study got referenced there. My own biased and deep-tissued take on this study is that it is the most important accidental agricultural discovery of these last fifty years as much because of "who done it" as because of what it says. And that, it must be said, in light of the fact that they discovered absolutely nothing new. The Marsden Project has establishment industrial-agriculturalists eating their own propaganda. The Marsden Project clearly and dramatically concludes that the craft of farming beats out the industrial model of agriculture It produces more food and fiber while improving the soil and requiring little or no chemical inputs - period - unless of course you want to go deeper in and say that it invites bio-diversity, more people on the land, improved water quality, revitalization of rural America and less hunger. Need I go on?

Karen Perry Stillerman, writing in the Union of Concerned Scientists Blog, http://blog.ucsusa.org/crop-rotation-generates-profits-without-pollution-or-what-agribusiness-doesnt-want-you-to-know/ does an admirable if limited job of presenting a suggestion of the implications of this study if not much on the application of same. Keep in mind that she as a science writer is speaking to industry NOT to farmers. Here's how she frames her discussion;

> *Substantial improvements in the environmental sustainability of agriculture are achievable now, without sacrificing food production or farmer livelihoods. When agrichemical inputs are completely eliminated, yield gaps may exist between conventional and alternative systems. However, such yield gaps may be overcome through the strategic application of very low inputs of agrichemicals in the context of more diverse cropping systems. Although maize is grown less frequently in the 3-yr and 4-yr rotations than in the 2-yr rotation, this will not compromise the ability of such systems to contribute to the global food supply, given the relatively low contribution of maize and soybean production to direct human consumption and the ability of livestock to consume small grains and forages. Through a balanced portfolio approach to agricultural sustainability, cropping system performance can be optimized in multiple dimensions, including food and biomass production, profit, energy use, pest management, and environmental impacts.*

What interests me more is how she then shifts slightly to qualify this study, if only peripherally, with a nod to how it might apply in the "real world".

Big Ag has worked hard for decades to instill a belief—in farmers, policymakers, and the public—that its chemical-intensive industrial farming methods are more productive than low-input methods, and more profitable for farmers. In recent years, study after study has cast doubt on this view, and now a team of government and university researchers has published perhaps the most compelling data yet showing that more sustainable farming systems can achieve similar or greater yields and profits, despite steep reductions in chemical inputs.

The so-called Marsden Farm study is a large-scale, long-term experiment conducted by researchers from the U.S. Department of Agriculture (USDA), the University of Minnesota, and Iowa State University. So no, these aren't California hippies or east coast elites. These folks know the dominant agricultural landscape of the Midwest—corn and soybeans. But they also want to better understand how systems that incorporate other crops, and even livestock, compare when performing head-to-head.

Keeping it simple (or not)
Over a period of nine years (2003-2011) on the Marsden Farm at Iowa State, the researchers replicated the conventional Midwestern farming system—a highly simplified rotation of corn and soybeans on the same fields on a two-year cycle, with copious additions of chemical fertilizers and herbicides. Alongside it, they grew two multi-crop alternatives: a 3-year rotation incorporating another grain (triticale or oats) plus a red clover cover crop, and a 4-year rotation that added alfalfa (a key livestock feed) into the mix.

I suspect we will be talking about the Marsden Project for a good long time. No doubt this is NOT what the USDA and its conglomerate brothers and sisters would prefer. It is our sincere wish that folks don't take the project findings and feel compelled to apply them as a direct simplified formula which encourages a modest return to crop rotation with a reliance on heavy chemicals and genetic engineering. That would be missing the point and the OPPORTUNITY. Some are already arguing that ANY return to grazing livestock on "crop" land would be a reversal because they would compact the soil and make no-till (that bizarre cousin to chemical warfare farming) more difficult. To reference Uncle Ephraim 'why the heck would we be afraid of straight ahead tillage when it is a proven tool for the very best of farming craft?'

This is the time to reinvite an abiding respect for the mysteries of life and how mixing and matching, overlapping and resting systems do give us our best farming future.

There are greater losses and most important lessons

Great Uncle Ephraim loved his Guernseys. They were his ladies. The herd dwindled as he aged but still, in his bachelor nineties, he never failed to milk the half dozen cows. The ritual reminded him of his entire farming history and kept him alive. His grandchildren had no interest in the farm or farming except that the land had come to be worth a great deal of money. Every morning, after milking Ephraim would drive the short distance to town and have coffee and eggs with an old friend and complain that no one was interested in what he knew.

As the family legend goes, Ephraim's grandchildren became more and more concerned for his comfort and safety. They couldn't understand how he at 90 plus years old could safely do the farm work and take care of the domestic duties himself. One day, on a visit, they found him out in the field working while the stove was accidently left on in the house. A family meeting resulted in the decision to move Ephraim, against his will, to a rest home. They had to secure a court order because he was completely against it. He argued "who will take care of the cows?" They promised him that the cows would be taken care of . He still resisted up until the orderlies arrived with the ambulance to forcibly take him away. Two days later, at the rest home, Ephraim's old breakfast buddy arrived for a visit and told of how the Guernseys had been hauled to the stockyard and sold for hamburger. The very next day 95 year old Ephraim died of unknown causes.

Epilogue: The family sold the farm and all the tools and divided the money convinced that they had done the right thing. The new owners of the farm ripped out the fences, bulldozed the house and barns and added the 160 acres to their 1,100 adjoining acres of corn and soybeans.

My job has always been to make sure that my Great Uncle Ephraim, the fierce plowman, always had someone hammering away at the need for better farming. Don't know that I have succeeded. I do know that events such as the Marsden Project offer the contradiction of a measure of good news weighed against the knowledge that it's late and that we lost so much treasure when we lost all of those Uncle Ephraims, all of those Fierce Plowmen who were waiting to lend us their secrets.

seventeen

in the midst

Necessary Questions
Acceptable Weapons
and Foisted Farming Futures

It had been freezing cold for weeks so when the temperature hovered around 34 degrees that January morning it felt almost spring-like. I crossed the fence and walked with axe to the pond edge to chop open a drinking hole for the 30 some head of horses and cattle. Been doing this each morning. That day the early morning sunlight set the frost layer on the pond surface to tiny sparkling crystals. The sharp axe chipped into the ice with the first stroke and I 'felt' a deep-throated rumble almost as though the air mumbled. Looking around for some evidence of the cause, perhaps a Rock Chuck clearing its throat or the garbled growl of a Badger or even antlers rubbing against a hollow dead juniper trunk, I saw nothing obvious and swung the axe again. This time the same noise was accompanied by a tearing preface sound and I noticed the travelling crack across the ice surface. Now it was obvious but no less wonderfully mysterious. Temperature from cold to less so, the weather had warmed sufficiently this last evening, and now with the sunlight teasing the surface of the ice-covered pond my axe blows were unlocking this taught layer enough to allow the slightest heave and yaw. The traveling cracks, disturbing the frost layer ever so lightly, looked all the world like tiny varmint trails skittering in a way to suggest God might be borrowing old Durer's etching needle and working to capture the moment in a drawing.

Terribly glad to see and feel these things, my cranky old soul holds for the moment gratitude for my/our humanity. In this weird time of toxic food, short sentences, glib response, false community, planetary uncertainty, biological collapse and decorated infertility I have been feeling longer bouts of *lost*. The global discourse that includes destructive farm monocultures, the prescribed 'inevitable' death of books and periodicals, the general acceptance of corporate rule, the constant polemic around what would constitute 'acceptable weaponry', the way we allow ourselves to be guided into tight little corners of lessened spirit, the way money feeds - and depends upon - deadly conflicts everywhere, the way youth is usurped and bled of its vital contribution by the postponements and intoxication which come of digital diversions, the way the fabric of life - the biology of our diversity - is invited piece by piece and everyday to leave. We humans are at risk of losing forever the spherical bearing of our very existence. Somewhere, once, it was written that humanity would/could come to find, at its core, a shaft running true within its bearing. We apparently got close but that was a century or more ago. We called it the golden age of farming. Since then it's been downhill. We've been running without the grease of determinant and determinating heritage and gratitude and that shaft has worn out of round, it wallows now within the loud and quaky bearing.

The best farmers are observers. We might be forgiven to think that they often look like they are waiting for something to happen when in fact they are in the very midst of everything - listening, feeling, smelling, looking. And in that way they are in command of the fullest range of questions. Because to be a farmer is to be in the midst. As such, farmers are also indicators of the near future of life. When things go bad for farmers in general, likely the same will soon be true for all others. But lest you think this be a tale of dire straights, remember that farmers are also heroically resilient. Give farmers half a chance and they will find their way to health and productivity and the world around them will join in the result. In that respect they are a lot like bees.

Our honey bees have always been indicators of the health of our biological world. They contribute to it, and with their bodies they monitor it. Our honey bees have also been tiny tireless workers assuring us, directly and indirectly, of a plentiful supply of healthy and various food. With nearly fifteen years of the mysterious Colony Collapse Disorder (CCD) responsible for the disappearance and death of a big portion of the world's bee population, we were given a tragic opportunity. All that death and destruction forced beekeepers, farmers and responsible scientists to look aggressively for a cause and a solution. But it was the beekeepers, in the midst of it all and aided by their observations, who began to command the fullest range of questions. They brought us to the point of discovery. We now know the definitive cause of the CCD problem are sub-lethal systemic pesticides, many of which are produced and sold by Bayer Crop Science LLC; slow acting poisons gradually destroying Bees (and likely humans as well) while, truth be known, our own protective agencies, our own government, rest complicit in the crime against life itself. It seems only the old and the young have the courage and stupidity to accuse the EPA, the FDA, the USDA, the US Congress, and the White House of accepting payment from chemical giants to continue the pesticide application. And the science? The Chemical giants provide the EPA with their own store-bought science.

It's been years now since France first outlawed these nasty chemicals, so we have the clear and exciting evidence - poison gone, bees have returned there, they have once again shown us their resilience. Nature is astounding.

Meanwhile in the US where we are still subjected to this terrible regime of sublethal seed treatments (primarily in large corn and soybean plantations), Colony Collapse Disorder continues its destructive cycle. The longer our environment is subjected to this biological warfare the less likely bees and other life forms will be able to reconstitute. It's about the money, lots of it, going to a few people in industry and government. The tradeoff? The global biodiversity and fertility that is every human being's birthright, life's spherical bearing itself is going away.

Its all a complicated business, and yet, I know, one which may bring some of us to yawns in the telling. Ironic if humans should find themselves bored while witnessing the slow death of nature and humanity on the planet. I guess those corporate heads who remain to the end will at least have the satisfaction of knowing they died holding all the money.

And its so darned easy to reverse. Just STOP USING THE POISON!

We live in an insane time of weird lop-sided priorities. While these agribusiness poisonings persist without legal restraint, local governments have moved to make it illegal to grow vegetables in the yards of private homes! In

2011, for example, a Michigan woman was threatened with three months in jail for refusing to remove a vegetable garden from her front yard while today in Florida, the city of Orlando threatens to fine people up to $500 a day if they insist on growing vegetables in their yards. I hear you saying "it can't be true?" I hear you asking "Why?" But those noises we make in quiet disbelief, not in suitable protest. As for the systemic pesticides issue? We aren't even asking questions.

It was late spring a year ago, when during the morning irrigation change I spotted old Windy, the thin 30 year old endurance Thorobred standing soaking in the sunlight. Old, so very old, but fortunate in her horse heaven existence here on the ranch.

That late afternoon, when I went back to check the level of the irrigation pond, I knew the molten shape spread so tight against the ground to be her, to be Windy dead. Can't say that I was surprised, she had had a long life. But I did not expect what closer inspection revealed. In a matter of hours or less the flesh had been stripped from her exposed ribcage, neck and face. And the tell tale signs pointed to cougar; more than one, because the striations in the flesh and the paw prints were of two distinctly different sizes - one full grown and the other a third the size. The head had obviously been attacked by a cub or cubs which pointed to the larger evidence being a she cat. I imagined a scene where the big cat tolerated the cubs at the "table" so long as they kept away from the larger prize. A terrible dramatic scene - but one purely of biological authorship. No corporate pockets were lined from the tragic death of this old horse. No sacred life mysteries were destroyed by plastic greed. Balance was maintained.

The story of Windy's death comes to us from the middle of life, not the end. Gruesome as it may seem it does run true in the bearing and in the hearing. The story of a massive chemical company inserting deadly product into the dead-end of industrial agriculture (with the two prime objectives of absolute control and high profits) is grim, horrific, and linear. Nowhere in the Bayer story is there any evidence of a respect for life let alone any loving regard.

As the electronic media world celebrates what it believes to be a new honest wholesale democratization of society, with a look-the-other-way attitude towards negativity and intellectual decay, it should be obvious how and why we have returned to a new dark age. We've given up on the observations of nature, terrible and sublime. We feel no need to tell the stories of nature's etching needle or an old horse's end. Lost and vanishing are the inherited narratives which grounded societies within societies. Narratives wrapped in ritual.

Poet Paul Hunter shared with me a tale of how the Navajo have long chosen to grandly celebrate the moment in each child's life when they first laugh; celebrate that moment and the person who is seen to have made it so. For, within the Navajo culture, this dramatic moment of passage - a baby's first laugh - is seen to be evidence that this new life has actually become a fully-formed human being. What a marvelous and grand excuse for a party! What a magical moment to share with family and community? Easy for me at least to feel its parallel tenor and tone with the best rituals of good farming - i.e. shared work-party meals - all contributing the silent aspects of a society's character. All very much "in the midst".

I recently learned of a trend in public education, given weight out of budgetary concerns as well as nods to contemporary relevancy. A trend which, in my less than important opinion, is evidence of the subtle and not so subtle ways we are making of ourselves an entity less human. Some school boards are arguing that it is time to do away with cursive writing in the curriculum. Do away with teaching the writing in long hand, along with penmanship - because, they say, the world has moved to keyboards and away from pens and pencils. (I guess that means they'll be saving all that money that has gone into team uniforms and helmets for varsity penmanship.)

Farmers know that children learn by being in the midst of life and living, by working to care for things, by feeling the doing. Penmanship is very much like that, exercises to see and feel the result of hand to eye coordination with enchanting result.

Ed Joseph has an important day job with the Department of Transportation, still he chooses to get up winter mornings at 3 am, harness a horse and use a feed sled, in the dark, to feed his small herd. This year his ten year old daughter, Natalie, has chosen to get up with Daddy each morning and go out, in the cold and dark, to help with those chores. Once the work is done, she goes back to bed and Daddy Ed goes to his highway job. With his crazy busy schedule, this is how she assures herself of "Daddy time". Each of us know that it is also shaping her character in most marvelous ways. She purely is in the midst, feeling the doing, and understanding completely the consequences.

Ed and Natalie (and Dena and Char's) situation is unique unto themselves. Most of us farmers have more conventionally hectic days. There is stress enough worrying about getting the field ready to plant, everything ready in good time. And then there's having the money for the seed and soil needs. Thankfully you were too busy with everything else to pay close attention to the actual growth, but its Tuesday and there it is, a fine crop growing straight, vibrant, and ahead of schedule. Luckily, while your mind was elsewhere, the crop was not visited by destructive forces be they pest, pestilence, weather extremes or visiting wildlife. You've made it to this point... Should you be thinking forward? Getting some things ready for harvest?

Preparations frequently bring reflection back to useful habits and purposeful respite. You pause and remember how those first days of plowing, ground slippery wet, furrows glistening, went so well that you figured plodding ahead with more days of nothing but plowing would be good, get it done while the plowing is easy. You'd forgotten how the cold spring sun still would bake that slippery wet furrow into concrete-like chunks. But, plowing done, you were reminded pretty quick with how rough a ride it was to harrow down, how hard the going was for the geldings, how it tore that one quarter crack open so bad that the Ted horse had to be taken off the work string for a month and a half.

Years ago you learned the value, with heavy clay soils, of plowing for half a day and then harrowing for the second half day, especially if that soil was slippery wet. These new implements, combining rollers and spring-tooth harrows, sure make a difference. But you were in a hurry this Spring and the plowing was going so smooth. Going back out each session to the same implement, same job, seemed a mental rest. Don't have to think too much, just go back out and do what you know to do. So you let slide your better self, that better-self earned over productive time, and you went for the easy shot only to be punished later. Learn from that, you tell yourself. Think now on what you need to have ready, physically and mentally for the coming harvest. Powerful analogy for all things 'life'. But perhaps not for everybody.

Many prefer to believe that we need not worry and prepare so much, that it will all come to us in the end. Like the musician in that children's parable of the grasshopper and the ant. Quite the perfect parable because it could be seen to prove both points. Ants succeeded through hard work and planning. Grasshopper, though threatened in the short while, made out like a bandit because the ants provided for him in the end. Lesson learned?

I certainly understand, at this late stage of life, how it is that the best laid plans and preparations, the most diligent of sustained efforts to stockpile and prepare, can shatter and blow away in the briefest of unexpected moments. While I would hold that all that work of the past does provide me with an undeniable inertia which keeps the motion forward, still it is far too easy to feel the threat of vulnerability. And for me the vulnerability lies with that terrible rhetorical question "must it all end here?" I confess that with the story of the "sub-lethal" pesticides, I am truly frightened of the answer.

So I go back to the child's first laugh and the violent passing of the old horse - neither of which, in our world of spherical bearing, mark beginning or end - both very much in the midst, both contributing to our fertile continuity, and I know that we might still triumphantly be in the magnificent midst of life.

eighteen

less talk more work

You pull hard against the load. Hard. Straining, leaning forward, all muscles targeted towards getting it to move, or to keeping it going. Then without warning, it snaps, something gives, and you feel yourself falling, face forward, out of control. Your feet can't feel the way to get out there ahead, as bracing. You 'know' you're gonna get hurt. This is not good. It's a form of dizzy helplessness that is frightening and disheartening. It literally and figuratively throws you off.

Now, imagine you are one of a team or set of horses or mules, in harness, doing a routine job you know and understand. You are pulling an implement, a plow. It slices through ground, lifting and rolling the dirt. This plow is mounted on wheels and your human partner sits above it, behind you, driving lines in hand, plow controls nearby. With no warning the pull tightens and you instinctively lean more effort into the collar as you feel held back. You push harder, your partners push harder, then something gives or snaps, all you know is that you are falling forward in panic. You and your teammates fall to your front knees, noses hit the ground, you jump up with a sense that something behind you is very wrong. You think about getting out of there. You might even wonder whether or not the teamster is still in charge.

Draft animals dislike being "cut loose" of a load while pulling. If it happens two or three times in a tight sequence, they might be expected to baulk when asked to pull that hard again. They might stop just as soon as the load stiffens, hoping to avoid the disconcerting "release".

These things came to mind recently while I was considering the new innovation designed into White Horse's Spring-assist Riding Plow. Canadian Kerry Smith made it possible for one of these new implements to make its way to our last Auction. Before our event I had spoken with Melvin of White Horse on the phone. He was deservedly excited about several things his Pennsylvania company have been able to develop. They had modified the basic design of the long-turning Kverneland plow bottom to reduce the drag and accomplish the same fine soil turning work. So now we have these bottoms manufactured stateside (instead of Europe). And they've set up shop to be able to reproduce plow points for most models of horse drawn plows. (Means that when you need an Oliver or John Deere plow part they will probably be able to help you.) But what seemed most intriguing in the moment was their new spring-assist sulky plow, designed so that when an obstruction was hit while plowing - the beam would release and swing back and up, allowing the plow point to clear the stone or root ball, and then as the plow moved forward, to reset itself. An amazing, sophisticated and useful innovation for those of us who plow with draft animals.

Day before our Auction, shortly after the plowing competition, Mike McIntosh and I had a very narrow window of time to try out the plow. The piece of ground available to us was hard and harder with mysterious subterranean aspects that felt like we were trying to plow over the buried cab roof of a 1949 Studebaker pickup. Mike and I are of that old school of thought when it comes to plow adjustment: change ONE thing - try it. If it works, bingo. If it don't, adjust ONE thing - try it again. Never ever adjust several things at once and especially not if you have lots of company most of which are convinced they are the only ones who understand the plow's problems. I'm reminded of that time Mike Atkins and I were going to fine-tune the then brand new prototype Pioneer foot-lift plow before the National Plow Championships in Ohio. It was raining. Mike had his superb three Belgians hitched. No sooner had we stuck the plow in the ground then 41 guys in bib-overalls all brandishing chrome crescent wrenches wafted down, ghost-like, from the surrounding tree tops and set in to twist and turn every moving part on that plow. No doubt some of them knew what they were doing or at least trying to do. But the others didn't and the result was a mess. Mike and I plowed off a ways in the rain and when we found ourselves alone, chuckled and adjusted ONE thing and tried it - that was the sequence until we got it just right.

The premise here, with adjusting plows, works in many walks and ways of life; often one small change has a big effect and we cannot always predict the outcome. Changes to two or three things in a complex or simple working system might throw understanding and fine-tuning to the winds. (Being in a hurry sometimes translates to being in reverse.) Take working animals for example. When mystery and confusion cloud efficiency and effectiveness, patience applied to the rule of single steps will always pay big dividends.

If you've been working horses for a good while you come to understand these things instinctively. There's a lot of shade-tree psychology employed these days when it comes to training animals to work in harness, some of it very good, a lot of it superfluous or even silly. I still lean towards what my mentors and life experience have taught me. Nothing but nothing replaces long days of actually doing the work. And that is true of both teamster and animal. Theories and arm chair admonitions (i.e. NEVER smoke a pipe around work horses and NEVER NEVER buckle your wingus before you tighten the goober-clamper), cloud the real objective. "Hey, let's just get this work done," he said. There's time enough Sunday afternoon over a glass of cider to wiggle our brains around all those 'should'a could'a s, though playing music would be more productive.

On the back side of that I can easily applaud the White Horse innovation with this fine plow because the intent is so very clear, "let's just get this plowing done".

So, in keeping with the suggestions of several of you readers, less talk here about philosophy and such - more of the nuts and bolts to getting the job done. Perhaps that way we can be better prepared when we're surprised because the load broke loose and threw us into the scary future.

nineteen

it is who we are

Out changing irrigation pipe on a crystal clear summer morning and I notice four buzzards circling over the woods a little to the north of me. One peels off the circle and heads my way, then floats on the current until it is just thirty feet above me banking into a small circle. Suddenly it dawns on me that we are looking right at each other, and that I am standing still as a corpse. All of a sudden, I jab my fist into the air at the scavenger bird and it jerks its head, falling backwards out of its circle and returns to join the other three off in the distance.

Reminds me of my dear departed buddy Bulldog Frasier. He was staying with us, as was his custom, during one of our auctions, he and his red heeler Stubby. Auction was done and we were having breakfast just before he was to leave to return to Montana. Fork in hand, no change to his tone he said "Boss, don't ever give up. No matter what, don't ever give up." Not long after that Bulldog passed away. But he never ever gave up. That was the sort of man he was.

Bulldog was a horseman, a farmer and a logger. He knew intimately what it meant, and what it took, to stay with the necessary work, day in and day out. He knew that there would be days when he could enjoy having laid up the crops, or having loaded out the last of the logs on a job, or selling a good team of horses he had raised and trained. But he knew just as certain, that every next day would have more chores needed doing. That he had signed on to a continuum.

With the difficulties we recently experienced, difficulties that arguably were not of our making, we almost lost the ultimate battle, because we almost allowed the difficulties to define us. But now, all of a sudden it would seem, we shake our fist at the buzzards, and we return to the real work at hand because the animals need fed, the crops need tending, the fence needs patching, the neighbors need our help, and family wants to be held and enjoyed. Those of us who are farmers know these things. It is who we are. And that distinction is incredibly important.

Though the evidence is to the contrary, in the world today society seems to have accepted without quarrel that the highest and best distinction for us all is our commonality. I disagree completely. I believe our highest and best distinction as human beings comes of our individuality, and of our separate and separated cultural distinctions. I believe completely that we as farmers are different from school teachers, I believe that carpenters are different from bankers, I believe that Japanese people are different from Sudanese people, that paupers are different from princes, and that thieves are different from honest folk. And I also believe that the lines of distinction are frequently fuzzy and blended. But that never lessens the defining facts of the distinctions, one from the other. And those distinctions, that variety, those various sets of working values give us our vitality and worth, they define us.

Visiting a Parable

Fifty-five years ago, in requisite summer bible school, my young brain took a bead on the story of the Tower of Babel. I found it fascinating even though I was too young to have any context to place it in, or against. As my remembered version of the story goes, way back sometime around the beginning of recorded history every one was of a kind, spoke the same language, ate the same foods, on and on. One nation, if you will. And the leaders, feeling like there wasn't much left to accomplish within their small and nearly perfect world, decided to have its peoples build a tower all the way up into the heavens, right up to God's front porch as it were. The project caused some discontent and, depending on your version, for whatever reason people fell upon each other in anger and argued until their languages separated in many dialects and people grew to hate one another just because it seemed the 'right' thing to do. (I do believe that is where we came up with the word 'Bable' as in nonsensical speech, a confusion of tongues at the Tower of Babel.) The Tower figuratively and literally came tumbling down and the small engineered and 'perfect' world became various, messy, large and far flung. I've always felt that the story contained a seed of the truth of natural design, that the 'world' sought and seeks its own balance in all things, definitely including the human species. And that balance begs for variety.

The Ranting Section

I have a storage closet in my brain, a space where I hang thoughts and ideas in a haphazard pattern that matches how these thinkings touch one another. The Tower of 'Babel' has come to hang in my brain with many thoughts centered on modern man and corporate rule. I 'feel' that corporate governance is very like the one world leadership of early Babel, believing that keeping everyone of a language and of a target (building the Tower) was the right thing to do, to demand. The board room needs to believe, in the measure they feel counts - the marketplace, that people in Uganda and Paraguay and Alabama and France and China are or will be all the same - they will eat the same foods, live in similar houses, visit with each other over the same social networking sites, use the same medical systems, worship in similar ways, get their news from the same sources. And the board rooms have decided that the Tower we are building is one of artificial life, that we will reach God when we no longer depend on the vagaries of nature for our food, environment, shelter, spirituality, class structures, and more. Genetic engineering, artifical intelligence, synthetic materials, 'virtual' realities, corporate funding and 'suspense' accounts will, they believe make it possible for all of us to travel back and forth from heaven on weekends in hybrid vehicles outfitted with talking computers which are capable of generating genetically-engineered snacks, beverages, and travel games. But something is going very wrong with this plan. People are fighting amongst themselves and reclaiming old ways, languages, heritage foods, craft-based skills of self-sufficiency, spirituality which is connected with nature, and an abiding disdain of usurpers, board members, pretenders, cyber mobs, stock brokers, bank owners and internet chatrooms. The construction of this new tower to artifical life is faltering. We are experiencing, in the wider world, a 'confusion of tongues.'

So I'm out baling hay and these scattered manic thoughts come charging in on me. What if several states in these United States actually succeed in seceding? What if Quebec becomes a nation unto itself? What if Mongolia pulls out of China? What if people around the world recognize that all of our governments are in states of advanced rot? What if the CEOs of large brokerage firms and big banks are actually charged with the felonies we all know they committed in recent years? What if Monsanto is tried in court for endangering biological life on the planet? What if college athletes are allowed to earn money? What if every elected official in every country has to pass a series of tests: lie detector, blood workup, spelling, arithmetic, criminal record search and financial disclosures? What if the Food and Drug Administration was subjected to an ethical and moral audit? What if people are allowed to grow real food? What if Facebook were restricted to those under the age of 10? What if people could milk their cows and then drink the milk, put manure on their fields as fertilizer, collect seeds from the plants they grow, allow their chickens and hogs to run out on pasture, plant fruits and vegetables in their yards, and be contemptuous of a justice system which has become contemptuous of them, a justice system which has forgotten what "justice" is? What if pigs grew wool?

Thunder and lightning and a whole mess of confusion, that's what it would mean. The Tower of artifical life tumbling down. It could also mean we find we're on our way back to a bigger world, one where fairness was/is a criteria and wealth a wide open measure. Because, let's face it, the race to all things "bigger" has granted license to stupidity, corruption, and police-state tactics. The breakup of governments, empires, corporations, and artificial integrations would, in the long run, be a mighty good thing. Allow tribes to be tribes, find the biologically defensible ways to keep them from fighting each other. Allow families to be small craft-based businesses and find ways to defend them from corporate marauders. Allow that even the smallest of us are innocent until proven guilty even if it erodes the court system's implied imperative to protect the property rights of the biggest. Does that make of me an anarchist? Don't think so. Think it makes of me just one individual who's tired of looking the other way.

Getting Personal

I have an Amish friend who once told me a joke that has served me mighty well over the years. He asked "What do you have when you have 5 female pigs and 5 male rabbits?" And you say, "I don't know what have I got?" And he answers "you're rich 'cause you got ten sows and bucks."

There it is in a nutshell. The answer to every measuring question any one has ever asked. We're rich, all of us, if only we could see the ten sows and bucks in our portfolio. Wealth a wide open measure.

The federal government, led by the current judicial system and executive branch, has come out these last few years completely on the side of industrial agriculture. They use food safety and cheap-food as thin and empty rationales for amplifying the protection, tax avoidance and welfare-funding of multi-national corporations. They'll tell you that concerns about manure as fertilizer, and fears of free ranging chickens and hogs, and health paranoia around raw milk are all about protecting the public from disease. What they don't tell you is that huge numbers of folk are getting sick every day in this country from the industrial food supply. To avoid that attention, the focus is directed at the little guys. So we are sometimes forbidden to take our work animals into market gardens because of the concern for ecoli bacteria. And, in a growing number of communities, we are fined for growing vegetables and fruits in our own front and back yards. And allowing pigs to graze on pastures is seen as disease-riddled barbarism? It is clear to me that now, more than at any other time in recent history, our government WANTS small farms to fail, wants them to go away, wants them relegated to some closet of historical relics. And they want it because the big corporations demand it be so. Monsanto, Cargill, ADM, Dow, and others want it because they know they cannot compete. They cannot sustainably produce food and fiber for the world. Small farms can and do.

Old Dan

Speaking of who we are, trying to fix my old John Deere A tractor I met a new friend named Dan. He's an old guy like me and lives out on the edge of town with acres of old farm equipment scattered about him. He parts out that equipment. Walking up to him for the first time, his face and hands greasy - big smile, easy manner - I could tell he was a rich man. Looking around at all the old equipment, I observed "What a wonderful sight. This is parts heaven." He just beamed. He's been gathering this stuff to himself for over 40 years, gathering the stuff and also the bits and pieces of knowledge that go with it.

I asked him if he had a radiator fan assembly for the Model A and he said "I think so, let's go have a look" so we threaded our way past balers, tractors, mowers, plows, swathers, and piles of parts til he got to the better part of a JD Model A tractor sans back wheels. The cowling was gone so we could easily see that the radiator and fan assembly were there. "It'll take some work, have to pull steering wheel and radiator to get to it. I might have one already out, let's go look." So we wandered at his pace, again meandering through his garden of parts debris. We came to an area which had no less than five or six more Model A's in various states of disassembly. No fans, except one from an older model that wasn't going to fit.

We got to talking about old machinery and I got the idea he was a protector or gate-keeper of obscure equipment knowledge. "Hey Dan, I've got a slow leak at an oil seal on a mower. Don't want to tear it down right now 'cuz I need to keep it in the field mowing. Any secrets?" "Yep", he said "mix a little clean grease in with that gear oil, might stop it up as the stuff mixes in. It thickens the oil."

"Another question for you Dan, my baler has an old Wisconsin engine on it and it's running rough and blowing bubbles in the filter bowl. My friend and I figure it might be the diaphragm for the fuel pump, do you have such things here?" "Nope, but you can get them from the manufacturer. Tell me more about that problem." I did and he says, "I don't think it's the diaphragm, could be the gasket for the filter bowl, take this one home and try it out. Anywhere that there is a hole or a poor fit you might be sucking air."

I knew that day that I had made the initial acquaintance of a treasure, a true treasure. Dan let me believe that if I ever had any other obscure farm implement parts question he'd have something useful to say. For years I have done such stuff for folks, mostly on horse drawn equipment. Don't always have an answer but have been at least able to frequently point people in the direction of a solution. Now here stood Dan Miller, no relation but every relation because he gave me the gift of understanding how knowledge shared is wealth. How knowing where to go for a possible answer is also wealth. He gave me a view of myself from the other side.

We didn't find a loose fan and shaft that would work and he understood I needed it soon. So we made a deal: he would come out that evening after supper and take the fan and shaft off that first tractor for me. He'd have it ready to pickup next day. I did some sneaky thinking out loud and discovered that Dan was partial to root beer. That I kept in my bonnet. So I left him thinking to myself that Dan and I spoke the same 'tongue,' there was no confusion.

Went back to the ranch and met up with my buddy Jon Peasley, who'd been out helping with hay. He had a brilliant idea to do a temporary fix on the tractor's radiator fan by strapping two large hose clamps on either side of the fan's hub. So we were able to get back to the field. But we'd also been trying to unravel a couple of mysteries surrounding my old JD 14T baler. There was that problem with the motor's fuel system sucking air, and there was a problem with the baler losing umpf when the windrows were big. Thought was that these problems were at least somewhat related. If the baler motor was not up to snuff it wouldn't be able to provide the power for the plunger stroke. So we fiddled with this and we fiddled with that, 'cause poor farmer that I am I cannot afford to go for fancy equipment. I've got to make the old stuff work. (Truth be known though, I am mighty partial to the old equipment, it suits me to a T.)

Jon Peasley said, "I think we need to tighten the drive belt." And I said, "But that seems contrary because right now it wants to stall out whenever we get into a big clump of clover. I think we need to get the motor to run better." We went back and forth on these questions til we decided to attack the problem one section at a time. We put the fuel pump back together, then dissassembled the carburator and cleaned it out. After replacing it we blew out all the copper fuel lines and accidently found that the line from the tank to the fuel pump had one end with no flare. It was sucking air in around the connector. We replaced the line and fiddled with the mixture and got that old Wisconsin engine to run smooth as silk. Still though, the baler was jamming up with the heavier hay. So we shortened the idler pulley arm til the belt was nice and tight and that baler ran so good that we started to snap shear pins on the fly wheel. More tweaking this way and that way and a little fussing here and a little wrench slapping there and we got that baler balanced so she ran like a young widow after a pie thief, with determination and pluck. It felt real good to get that old equipment humming. Couldn't have done it without Jon's help because we were speaking the same language all the way.

Next morning I took Jon Peasley with me to get the tractor fan from

Dan. I remembered to pick up a big jug of root beer (his brand). When we got there, there were several guys waiting to be helped so we just wandered the garden of farm equipment parts. Peasley was having fun, he's got partial tractors of his own. Dan finished loading the old carcass of an International 460 on a flatbed trailer then drove his rusty old forklift our way. It was morning but already 90 degrees. When I handed Dan his jug of root beer he smiled big as a chrome radiator and said "Huh? You done what you said you would." He made the observation with just the gentlest note of surprise mixed with certain gratitude. No confusion of tongues here. No tower building. Just neighborly.

Finishing Up

All this stuff about folks sharing baler balance and farm equipment repair feels comfortably like the working reality of farming. Nothing about it is artificial. So much of it falls in the category of acquired and inherited knowledge and experience. Not so out there in the world of big business and big government, that world I alluded to when I was talking about today's effort to build a Tower to artifical intelligence. The quick and ready approach to most of the bigger problems of the world seems to involve either throwing lots of money at it or sending it back for further study. Take the problems with our environment: most school children know the scope of the problem and have a working grasp of what needs doing. Not big government and big business - they want more study on the question. Why? Because right now the perception is that anything we might do to correct the negative impact of man on the environment will cause big business to lose profits.

But you don't want to hear that because it's old news, old arguments. It's boring. What you want is the unusual, the dramatic, the quirky, the impossible - like woolly pigs. And by woolly I AM talking about pigs that grow a coat of curly hair what looks like wool, just like sheep. Would that catch your attention?

Most of us sit right here, certain that we have a clear sense of what the world offers up. We know about which vegetables can be found. We know about predictable weather patterns. We know what pigs should look like. We know, or think we do, about right and wrong. Or at least we know what is acceptable behaviour and acceptable fact. But do we really?

We know, deep down, that it is not good to mess with Mother Nature. We know how powerful she often is, especially when she hurls massive storms our way. Yet we continue to collectively look the other way when it comes to environmental pollution by industry, calculated and deliberate genetic mutations of life forms by science and industry, and geologically destructive drilling and mining practices. All examples of man messing with nature.

We look the other way because of the perceived tradeoffs. WE figure this is acceptable behaviour because WE "need" the industrial profits, the synthetic foods, and the oil and gas. When we are collectively given a choice, as in 'do we starve or produce more genetically engineered foods'? we jump to respond without questioning the starting premise - the one which infers that without genetic engineering humankind cannot produce enough food.

But all of that is when we think and talk from that position of the larger WE, the world universe WE, the tower-building WE. When, instead, we think, talk and act from our intimate community selves - as computer programmers, as citizens of Indiana or Uruguay, as the local union chapter of steam fitters, as truck drivers, as expatriate Russians, as Vietnamese school children, as Bayou fishermen, as black French-speaking Jamaicans, as small farmers - we do so with identity in hand, we do so feeling the need to hold and protect that which we value. Protecting our environment is like doing the chores every day, we do it because it is who we are.

It is NOT a small world, it is a BIG world, as wide and various as you can possibly imagine. We are not alone. When we feel ourselves shut down, crowded by worry and a sense of failure, it would serve us well to remember Bulldog's admonition, "Boss, never give up, no matter what, never give up." Anyway, how could we? Who would put up the hay? Who would unharness the team? Who would milk the cows? Who would wax the cheese? Who would feed those woolly pigs? It's got to be us, after all it is who we are.

twenty

a mulch of time

When I was young I wanted to be older. Seemed clear that older folks got respect, access to good stuff, and authority. Now I'm older and I see that respect is up in the air, access to good stuff is sorely over-rated and authority means nothing to the long haul. As an old man I replace authority with community. Now I know that I need my family and friends around me.

When I was a young man at one of my first ranch management jobs I had to take care of a couple of beehives. It was a case of learning to swim by being thrown in the pool. I was lucky in two ways, first bees (or insects for that matter) just did not worry me. And second, I had a friend, Wilbur Long, an old guy who was a beekeeper, he gave me some valuable insights. I enjoyed the bees. That surprised me. When I left that job, the bees stayed behind. For decades I wanted to have bees of my own. That happened this last Spring. Two hives of Italian bees; one standard 10 frame hive and one 8 frame English garden hive now live with us, stationed out on the edge of our pond sheltered by a grove of Cottonwood trees. They spent the whole summer with their backs to 70 acres of Clover and Alfalfa. It was easy to see they were happy.

Once the hives were established, my wife and I would go out and just watch those lovely bees coming and going. Impossible to fully describe the fascination and comfort, yes comfort, that we get from watching those golden buzzers. And then, seeing the incredible effect the bees had on all manner of flowering plants and trees I felt silly that we had waited so long to get restarted with them. All that time lost, but then...

One tangible way time exists for me is that I see it in the aggragate mulch or ground cover of my life, a blanket of experiences that keeps getting added to. That mulch shades the basic inescapables, the nasty and mortal shape of me. That mulch, that blanket is something I can measure. That mulch conserves my spiritual moisture and helps me to continue growing. Now this year's experience with bees goes into that measure. I can say these sorts of silly things because I am old. When I was younger I couldn't get away with it.

Some of you out there know what I'm talking about because you too have gotten old (or older). I have to ask. Do you enjoy it like I do? Being old I mean. I wouldn't trade my age, experience, gratitudes, or my mastery for anything youth has to offer. As for these stumblings, forgetfulness, this occasional crankiness, the soreness, stiffness, the limited movement, all of this aging crap, it pales by comparison to what young people have to suffer through. But, appropriately, it is us old people who look like we're having difficulty. Someone asked, "How do you get through the days, old as you are?" So I suggested that you do it one day at a time. Today you let a cold fall full moon slap you gently as you return from the barn chores. Brush the horse hairs off your vest and untie your boots. Kiss whoever met you at the door and key up a gypsy guitar, say Stephan Wremble playing Bistro Fada. Feel the inside air fill with lamb, raisin and tomato curry smells - or an apple cinnamon pie. Let the day's accomplishments pile slowly on your brain, in that comfort zone you save for those best over-the-shoulder looks - it's for those times when you seeing what you've done makes you feel good. Well yes, there is stiffness, soreness, and fatigue too - yes. But don't let yourself follow the logic. It is far too simple to allow yourself to think that it used to be so much easier. Perhaps it did, but the gratitudes were far thinner. Of an evening as a youngster you seldom found yourself ahead in the count at days end. These days, old man, with half a moment to yourself you are overwhelmed with how fortunate you have been in this lived life. (And now the bees are back it will only get better.)

Later the extended reverie may pry open those moments when we find the nonsense questions take over.

"What ever brought me to choose this farming life?"

"Where has the time gone?"

"Seems before there were always folks around to help."

Tweny-five years ago, around our remote ranch, surrounded as we are by forest service ground, every spring Paul Reuter ran two bands of sheep, about 2,000 in each group attended by a Peruvian sheepherder with two predator control dogs, one horse and an old rifle. Roberto Alania was one of those sheepherders and he became a good friend. When we first bought the ranch there was a dirt trail we drove to get to town, all dirt and bumps. I remember the first time, rounding the bend at Squaw Flat, when we came upon a wispy dust cloud populated by slowly wandering sheep, what seemed an endless flock. In that dust, sunlight forcing its way through, the outlines of the sheep seemed ghostly. I was so taken by that sight that it became one of my paintings. But that's jumping the gun so to speak.

Reuter came to me one day and introduced himself saying he had a broken frame member on the old travel trailer one of his herders used. He needed to pull it to the next grazing site and was wondering if I had a welder he could borrow. I said yes and that began a shared relationship that lasted for all of the years during which Paul was permitted by the Forest Service to continue grazing. Seemed he had been driving 24 miles round-trip to draw water out of the lake into his tank truck. That water was then taken to the portable troughs for the moving sheep. I told him he could get his water from us and save many miles. He was most appreciative, keeping our freezer full of lamb.

Why am I talking about this here, and now? Because as I wander over the years of adventures we've had on this farm/ranch and allow the reverie to go into the cracks and creases of the stories, I realize how rich and varied our lives have been, and that all of it goes into the mix. All of it answers those last two questions, where has the time gone and who used to be part of our daily lives?

As we would encounter the shepherd camps, we'd occasionally see their hobbled saddle horses. After a couple of years passed, helping Paul with water and tools, one day we found the familiar herder's saddle horse standing adjacent to our stud pen. She had come to visit old Abe, our Belgian stallion, walking in short hobbled steps well over a mile just to let her beauty be seen. I called Paul and he let the herder know where to find his mare. That was the day we first met Roberto Alania, one of the herders. He is a highland Peruvian of purest Incan descent who spoke, at that time, in halting broken English. I, with my similarly restricted Spanish, found ways to talk with him. Those days my brother Tony helped us on the ranch and lived in our second cabin. He and Roberto became close friends, sharing meals. The Forest Service made a big mistake when it arbitrarily decided to end the hundred plus year history of sheep in our country by revoking Reuter's grazing permit. They know it now, for the land has shown them over time what a beneficial tool that grazing cycle had been. Our locale has many features which remind Roberto of Andean territories in South America, as we are up against the dramatic Cascades mountains. He fell in love with the country. My brother had left by then to care for our folks in Florida when Roberto came to me looking for work on the ranch. He missed the country. He worked for us off and on as we could afford to hire him, and he became a good and trusted friend. He lives and works now elsewhere but we see him occasionally and reminisce about those times. He and the story he shared with us as working companion is but one of hundreds that have made up for us, to this date, the storyline of our farming life. And for this essay I offer that there is no way anyone could have factored in such an exotic and elegant relationship when, as youngsters, we laid out our plans for what sort of life we would want out of farming. Nor could we have guessed that most of these unexpected side stories would contribute so much to the reward that has been this life.

But to shift sideways for a second: There are two conversations here and perhaps they will join at some point. One is how the passage of time, and a sense of its inevitability, sits atop our lives as farmers and would be farmers. Second, from a societal or cultural perspective, how time does drag along those geo-political, technological, religious, and slap-happy diversional changes that would affect us all if we let them. And in that "If" is the waft of choice that makes inevitability a joke. In other words inside of all of this are choices, personal and collective. Obvious, but deeper yet are the determining aspects around who we allow to decide for us, collectively and personally.

Some would say; "In the beginning, we did it for the kids. Now they're gone so I suppose we're doing it for ourselves. Sometimes I like to think we're doing some wider good by sticking with farming, but its hard to hold on to that with all that's happening in the world."

Others might say; "I did it for myself because it represented the lifestyle I wanted. Then there were two of us, so I suppose it became what we wanted. Then came the kids and it was difficult because we needed to generate more money. Now the kids are gone and we're finding we can enjoy it more."

And still others would say; "We started out as a family and there were many of us. We were all excited and everyone contributed to the work. It was hard but it was worth it, and having many hands meant we could avoid the big machinery. But now the kids have left to other lives and there are just two of us. We wanted to stick with the work horses so we decided that it meant going to more sophisticated implements, tools that replaced the extra hands. It's different now, we are no less committed but I have to say that there is a very different flavor to the days. Perhaps it is that there is less evidence of craft and more evidence of production."

And a sad few would offer that their adventure was stripped from them by difficult realities including a string of bad luck, regulations that crippled, competition that rolled over them, until - failed dreams in hand, devastated spirits pocketed - they left.

We were young once, many of us. And in those days had drawn a bead on what we would become, where we would settle, what lives we might lead. Reading this publication it might be safe to assume that your own scenario involved a small farm of some sort, or an adventure with draft animals, forgivably romantic or nostalgic while camouflaging the real core - a desperate need to feel connected to a living life with a reason to be. And then we got old. But in-between those beginnings and now we lived through the time. Did we live the time? Did we spend our lives wisely, or at all? Or did we tuck our lives away, waiting for a future when we'd get around to that dream of right livelihood?

Some of us are young right now, and do not feel it, refuse to allow it to define us, insist that the distinction is a distraction at best. This publication represents for us - the youthful - a tapestry of mistakes and possibilities, an indication that others, older and without the benefit of our perspectives, advantages, and energy, were able to make a go of a handcrafted rural existence. This publication is to us - the youthful - like a limited guidebook for small farming, limited because it insists that the cyber universe is irrelevant and that dues must be paid. We the youthful know that the world is littered with people dumb enough to pay for knowledge and experience and that this is not going to happen to us. Somewhere in all of that haphazard choice young folks seem to have lost sight of the fact that they need extended community and that most definitely includes older folks.

Though it is nowhere near as simple as described there is nevertheless a canyon or canyons that separate the ages. It does not need to be that way. And the separation is a loss for humanity. In a recent telephone conversation a subscriber told me that she wished she could write a story, for the journal, of and about the old gentleman whose farm she manages. But she couldn't do that because if she did she would have to be honest about their great disagreements. She said that when he passes, if the farm should transfer to her, there would have to be big changes. When I expressed my concern that this was unfortunate, and that both she and he would benefit from just such an open transgression of confidences, she said that she didn't want to be misunderstood, that this old man was an incredible resource and inspiration. I wondered out loud at her loss should she never hear his response to her criticisms. She knows that I insist that the story be written now and published, here or elsewhere, making sure that he read it. I said it but I know well that it won't happen. This breakdown is representative of the wider disconnect between the ages.

Because what we are talking about are the voyages we took, take and want to take through a living life of farming. There is more commonality in that then there are differences between the youthful and the mature. Respect as artifice is nearly worthless, respect earned by shared experience and co-measure is priceless. And respect does not mean, for me, a polite regard. Respect must also embrace the most hurtful of intimate differences if it is to be all that it may be.

In the interests of full disclosure I offer that I am sixty-six years old and began farming four and a half decades ago as a complete novice. I remember in the beginning that I felt disregarded and ridiculed by many of the old-timers I admired. I do not recall any of us as beginners ever disregarding, or politely dismissing the old-timers. My memory frequently belongs to me, which is to say that I mold it to suit my viewpoints. Maybe we were mean to our elders and I choose to forget that? But again, does this mean that perspectives are hostage to view point and that view point follows us always, changing as we do?

All of that, whichever of the myriad scripts fit, came of wider environs, the world around us. And those environs, political, economic, and cultural swirled and slopped back and forth bringing new pressures and colors to each of us. Whether we felt pushed to move on to a farm because of the threat of Atomic bombs, or riots in cities, or libertine excess, or cyber bullying, or pollution, or inviting cavalcades of like minds, or the crush of violence - each of us married our personal wishes to the times we lived in and came up with our own unique formula for living and working.

There is a ridiculous saying that goes something like this; "the more things change the more they stay the same." Social engineers like to point to how fashions seem to repeat themselves over time. And all of that I believe comes because we humans want to believe there is a pattern in all this, that there will be some predictability. But alas, such is not the case, and our present time is a prime example. We've never been here before, and it is most frightening.

I might suggest, strongly, that the grotesque mistakes of our Orwellian US government around alliances with corporate agriculture running parallel to the absurd rush by China to artificially spur its middle class by destroying its small farmers will coincide to create a devastating global economic storm that could and will force we small farmers on the ground to greater clandestine effort or career change.

It is only a matter of time before the oligarchs relegate natural foods, the handmade, craftsmanship, and independence of spirit to the dust bin of human adventure by outlawing any production outside of industrial control. It has already started with efforts afoot to make it illegal to allow chickens and pigs on pasture, to further criminalize raw milk, to jail people who insist on violating local ordinances to grow vegetables in their yards, to brand fertilizing manures as toxic substances, and to increase taxes and fees on proof of origin.

As an old warthog rooting around in the mulch 'neath the hourglass, I for one say it is time for civil disobedience when it comes to our right to farm and our choice for right livelihood. And, though I certainly do not encourage anyone else to follow, I intend to keep information about what I grow, how I grow it, where I grow it, how I harvest it, how and where I sell it and what I call myself and my products to myself. I am done sharing information with local, state and federal governments I can no longer trust. Difficult as it may be it is either that or trade in my farming tools for a job at Facebook calibrating insincerities. I expect that some of the young 'invincibles" will find my thoughts on this to be typically old and cranky, perhaps even paranoid. That worries me. Not for my own sake but because ahead is a minefield for which some of us old cranks just might have a map. I wake up from bad dreams hearing youngsters scream at me "why didn't you tell me they could take away my farm?"

We chose farming, choose farming, for as many reasons as there are ones amongst us. And today, though many young people seem drawn to a single tight model of farming adventure, tied by a counter-cultural fiat to the idea of market gardening primarily for vegetable and fruit production, far and away the majority of newcomers are distancing themselves from the corporate model of industrial systems agriculture. When I was young the magnetic pull was towards the notion of the old general farm with mixed crop and livestock production. Grazing and crop rotations, fallowing, the growing of livestock feeds to keep operations as self sufficient as possible, the application of both animal and green manures for fertilizer, all these things and more were central to the notion of a well-rounded and sustainable farming venture. I worry that having so many new farms insular by target and scope to just gardening becomes an imbalance going forward, one which will not be so sustainable. But I most certainly could be wrong and the best evidence of that has been four thousand years of Chinese agriculture (Read *Farmers of Forty Centuries* by F.H. King) wherein that model of local market gardening, coupled in many cases with orcharding and aquaculture, fed the world's largest population with legendary variety and great musical sufficiency (Read *The Last Chinese Chef* by Nicole Mones). Now that Chinese model is doomed to swift and certain death by government edict as tens of millions of small farmers and gardeners are pushed off those centuries old farms to move to urban areas. So how will China feed its people today and tomorow? Why, with bioengineering and chemical inputs as the all-powerful Americans have, of course. And if that fails China will use its vast stores of cash to purchase food from Argentina, Australia, Iowa, Africa, and Europe. They say "don't worry about us, we'll be just fine" but they won't. They will cave in upon themselves to civil unrest, disease and pestilence. The boards of Monsanto and friends say, "if that happens they brought it on themselves, it is not our problem." They too are wrong because it will also bring the great dragons of industrial agriculture, Monsanto and friends, financial ruin.

In the late fifties, when I was in my early teens, I distinctly remember folks talking about the cruelty of time's passage, mostly from the standpoint of a lament for the good old days. Later I would realize how scatter-shot that lament was because for my family and friends the good old days might ironically include world war one and/or two, the two terrible depressions, several recessions, and Jim Crow south - all of it horrific. How malleable is our nostalgia for the good ol' days. And what value has the lament? Might I suggest it has great value? For in that wandering wonder after what we have lost is the buried collective certainty of how things should be, could be, would be only if....

And now we must prepare ourselves again for everything to change dramatically, perhaps for a very long time. A big change. We saw changes over the last 150 years with the great events and shifts in our culture: wars, depressions, industrial and informational jumps, political storms. But, I suspect nothing so dramatic as what is coming now. Good chance that many of the societal elements we've come to accept over decades are going to go out the proverbial window: elements like how we gather, where we gather, what we share at primal levels - meals, civic meetings, religious services, cinema/entertainment, ceremonies, work parties, and rites of passage. Hundred years ago country folk might have been expected of a pleasant evening to sit together on a porch and sing or play music (a few still do), fifty years ago they'd maybe sit around in a living room watching television or attending a movie together, today...?

When you're thirty something you might be expected to make some pertinent observation or critique of coming social change feeling, reasonably, that we still have time to embrace it or reject it. When you're sixty or seventy and big change comes at you, measurements are made to determine whether or not you can stay put, hold to the old status quo and keep your comforts and rhythm and that's because, number one, you probably won't be around to deal with the outcome in the distant future and, number two, you probably don't have the political (or societal) capital to enter in to any fight to avoid the coming change. (Old folks today, just as with yesterday, are of questionable value to the young invincibles who sincerely believe they are creating everything right now and for the first time.)

The young work towards their goals with a short list of intense urgencies, very little experience and scratchy applicable evidence to brace their gamble. Used to be people in such circumstances would look to sources of shared information that would provide the sort of bracing that 'community' provides, such as the anecdotal evidence that others are setting out to do the same things, are succeeding at the same adventures, are struggling but no less committed to the same adventures. That's where this publication sat for decades, a resource and access point for community connection. Unfortunately that's not so much the case these days. And its not just that young people are getting what they need elsewhere, it is also because they are impatient to a self-destructive degree. Or is that the observation of an old person who simply doesn't get it? I see the readership of *Small Farmer's Journal*, young and old, now being more individualistic and less community oriented. Is it true? Or is it to be expected of an old wart hog like me?

The old discover that many of those goals they had when young were right-on target, but that the trajectory frequently threw them off in tangents. For some of the more fortunate, the conceit that would have them in "control" of their own lives gave permission to write and rewrite their own histories as they went along. 'Fortunate' because these folks, folks like me, can make like everything is as they planned, a positive accounting that fluffs along the next steps. Of course this is also a dangerous game that sets us up for terrible dilusion and then regret. But better that we had 'helped' ourselves along through the forest of life with assurances that all is as it should be, than to be people of that modern ilk for whom the forest of life is synthetic/artificial. Isn't it far better to discover that a real person you believed was your friend was actually a manipulative weasal rather than to find that your friend the 'warranteed' person never existed at all, was the invention of software engineers? Isn't it better to have participated in your life rather than to be a dweeb addicted to the video game that is social networking? Isn't it better to have actually assisted at the birthing of a lamb even if that lamb didn't make it?

Commerce says we are shaped by our acquisitiveness, inside we know very early that we are shaped by our affinities. And early on, these crazy days, people find themselves having to choose between those instinctual affinities and the constant cavalcade of extreme and bizarre virtual realities designed to trigger the basest of human crudity. It is delightfully intriguing how the life-long pursuit of those things and aspects we have always felt an affinity for delivers us experience that builds our own individual skill-set.

Early we <u>suspect</u> that security must be considered. Later we <u>know</u> that security must be considered. When do we learn that personal security comes from our skills-set and mind-set married to community and family not from gadgets, trinkets, arms, or money or power or gaming? Building our own mulch layer of experience and thereby skills delivers us to that wonderful comfort of knowing who we are and what we are capable of - individually and collectively. Time isn't the thing that delivers this, there are many old folks without these strengths and this self knowledge. Time isn't a passage though it does pass. The cummulative - that's the secret. The slow deliberate accumulation of deepest experience, skills, and relationships. We are the nuclei of fertile hope manifest, properly searching to accumulate the golden intangibles. Time is either the corrosive or the ultimate filter. Of course, who am I to say any of this?

I want to think I am you at a different time and in a different place.

"I know. But I have no evidence."
Italian poet Pier Paolo Pasolini

It's been 38 years that I have functioned as the editor of *Small Farmer's Journal*. Took that long to realize that for many years now some people have been pretending to themselves that they read and appreciate my written contributions. Others who do read those words often find them troublesome. Some even hang on hoping that I (and we) will return to a quiet tone and comforting content they remember from years past, a tone that arguably never existed.

So here I am remeasuring the audience for the SFJ and these words. And through the phenomenal correspondence and communications which come our way, I continued to be reminded that our audience is our community and it is close-held and immediate, its not in the wide measure, its in the one-on-ones, in the small groups and families.

Live long enough and you find yourself asking " do I have anything of value to say anymore?" I know the answer to that one, of course we have things of value to say. But the question is a distraction, the real question should be live long enough and will anyone listen to what you have to say? And why should they?

Be young enough, hip enough, arrogant enough to embrace your arrogance and you will run straight into that wall of fate without a trace of a smile and even less gratitude. Buy into the board room hierarchies of who decides what's best for the rest of us and assign yourself a seat at that table and you do so at your own risk for the end you guarantee goes against biological life - and remember you are a piece of that biology - you go against yourself.

So why write a silly magazine about small farms?

Because it is more important with each passing day.

As the old wart hog I've become I think the answer is that I must embrace my new found crankiness and go slowly through my remaining time in this heavily-mulched forest of humanity mumbling my own versions of truth as I root for and dig for useful reason. Like a stone-deaf Beethoven I need to write these words to "hear" them inside myself and use all the remaining hours of each day to farm and paint. For society's kangaroo courts, be they cyber or brick and murder, are a total waste of time, waste of honor, of spirit and of human dignity.

True earned community I believe in but sadly I am less interested in the shattered random shapes we take as groups of humans, groupings owing more to commerce and class pretense than to shared values and experiences. I am, though, still interested in how times do shape us. Great farmer/artists like Shakespeare, DaVinci, Jefferson, Lincoln, Cassat, Cezanne, Ronald Coleman, Albert Einstein, Edward Hopper, Duke Ellington, Lightning Hopkins, Joni Mitchell, Bill Evans, Nelson Mandela, Albert Camus, Red Skelton, Winton Marsalis and Johnny Cash were and are all noble products of their times NOT products of group membership. And they are all stellar pieces of the extended true community of mankind, a community which never abided with the notion of card-carrying 'membership' and committee rule.

We are born with all the answers, it is embedded within our biology. What we don't know, in the beginning, are the questions. That comes as we are served up a smorgasbord of curiosities leading to personal appetites. The people around us, our class-conscious society, then demand of us that we either develop or accept justifications for our appetites and in that process society messes with any ready sighting of life's true questions. If we are fortunate the questions will come to us and give weight to those answers we've long held.

Genetically engineered life forms are not born with the answers because they are born from a dissecting, melting and melding of genetic codes that destroys all the maps back to what makes them and who made them. These plastic-mutated psuedo-life forms have zero implicit or embedded regard for any true life.

Back at the beginning of this essay I mentioned happening upon those sheep blended into the dusty landscape. I did a painting of that scene. As I began, sketching with my brush the outlines of the flock, Kristi, my wife, saw this and remarked on its instant impact. I stopped the painting, turned it to the wall, and waited knowing from experience that with time I might be able to see it fresh and perhaps see what made that impression on her. She was right. That painting which I now call "Sheep Passed This Way" has become a testament to the powerful value of individual memory and how the right questions are always around us, nearby yet sometimes turned to the wall. We just need loved ones, family, friends, our community to help us to see.

ORLOFF STALLION AND SHETLAND PONY.

twenty one

charting

When Bud Dimick was 80 and I was 40 he came out to the ranch to help me mow hay, brought his own team and mower. That first morning he stood his team in one double tie stall and I had mine in the adjacent one. I noticed that he was having a bit of trouble lifting the harness up to his gelding's withers. "Here," I offered, "let me help you with that." He turned, dropped the harness to the floor and shoved me against the wall. "Don't ever do that again!" He said, "I do my own harnessing."

We went on to mow that year and for a couple of years after. He was a good friend and, in my eyes, a real giant of a man. Last time I actually saw Bud drive his horses he was 94 and in a parade. Bud passed a couple of days after his 105th birthday party at which he fed himself and visited with family and friends.

Sometimes, when we think hard about a subject, images and stories from our past seem to jump up and say 'here's an example' even though it may not seem so on first look. I've been thinking alot about how to get the right sort of folks personally, intimately interested in the future of the craft of farming. So stories like the one I just offered jump up and say 'here's an example' and I am stuck with figuring out how that may be. Perhaps it is so obvious it is difficult to see.

The opening vignette speaks to what it took to be long-lived and heroic Bud Dimick, a self-made man, a blacksmith, wheelwright, race horse trainer, farmer, wagon-builder and teamster - every one of those vocations/crafts he perfected by his own measure. He was tough - on himself.

We hear it. "I want to be a farmer, a true farmer, but I don't know where to go for land, for money, to learn. Who will help me?"

Doesn't sound particularly tough-minded. When I hear it expressed this way I want to turn it around to: "Do you have what it takes to be a true farmer? Where might you go to learn? What do you have to offer in trade for that learning. Are you gonna stick with it?" And that leads me to think about the place of elders. Maybe one of our jobs is to thresh out the candidates for this 'tough' life? Maybe, not every person has what it takes to make a go of a life as an actual farmer?

Bud didn't come all the way out to my ranch to help me because he wanted to help me. It was something else. On his small plot in Madras, Oregon, he didn't have a place to mow and he loved to mow. He appreciated the moronic repetition, going round and round a big field to harvest a crop. He knew what it would do to the conditioning and responsiveness of his horses. He appreciated how it made him feel. And he wanted the simple direct challenge of measuring his own work with the work of the other teamsters also mowing. In fact, on one occasion, while he and I were both mowing a forty acre field, I with Cali and Lana and he with his big roans, I with my number 9 mower he with his beloved number 7, he finished that first day with a surly nod my way and said, 'guess you just got better horses than I do'. He was a taciturn fellow, not one for lots of words. That statement was loaded with everything from self-criticism to compliment to coaching tip - he was saying in his own way that I was tough enough. Now, a quarter century later I come to wonder if toughness is enough, especially for us old guys. Perhaps without him knowing it, Bud was threshing a candidate, me.

How do we make new farmers? Or is that our responsibility? I was well along with this writing when we received an SFJ article submission entitled "Heritage" from Pennsylvania's Vastine family. That essay offered a clear case of the classic way young peoplle are brought along by a vibrant family tradition. This has become more rare with the passage of time. Today people who are one, two or three generations removed from farming find themselves attracted to the life but without those traditions that leaven. Are new farmers of the actual brand the product of society or purely self-made? And why do we ask this of society in general instead of asking it from the community of actual farmers? If farming is an art, and I believe that it is, where are the similarities in humanity's crush?

Humanity is in a world of hurt. Some folks think they know right where the problem is, and they'll tell you in a quick minute. I'm one who's come to believe that there is no single problem with humanity (or society). I believe that, though each problem is certainly bad enough, when you combine them they add up to more than the sum. In medicine they might refer to this as the 'double crush' with lots of incidents demonstrating how dramatic improvements can be made when even just one of the problems in the mix is corrected. Humanity has come to disregard tradition, to segment and target education, to turn a blind eye to nature's rhythms - all of this while feeding war, breeding pestilence, and pointing entertainment's cameras on human suffering. I suspect that if we were to rebuild tradition we might see a natural move to lessen poverty and hunger. A stretch?

True masters of the arts are the purest gold of human society. The arts include the obvious list (music, dance, theater, literature etc) but of course that list must also include any long developed useful human endeavor which combines creativity and craft such as actual farming and the shoemaker's art among others.

Alchemy by historical definition is the faux science / magical ability to convert base materials to gold. It doesn't need to be a stretch to see that taking the raw clay of undedicated humans and bringing them along to the pure gold of artistic mastery requires alchemy, the illusive recipes of which should be of the highest value to all of human history.

Definitions to think on:
Legacy: the cheap and quick computer definitions include: a gift of personal property by will, a longer definition might include a measurement of what you MIGHT have to hand off to future generations.

Heritage: practices handed down from the past by tradition.

Tradition: the sum or range of what has been perceived, discovered or learned. A specific practice of long standing, an inherited pattern of thought or action.

What constitutes tradition when it comes to heritage and legacy?

What is today's view of tradition, heritage and legacy?

An apprentice works for an expert to learn a trade, or at least that is as it should be.

Boot Maker's Turn

George Zierman is my friend. His business goes by "George's Boots". He makes exceptional boots. I wear them. George is a master craftsman and artist of the first order. He makes each individual boot to fit that foot. And he takes his opportunity as a responsibility. He wants to have the boots help to make the wearer more comfortable. George has been a longtime reader of *Small Farmer's Journal*. We met in person back at one of our early auction events in the 80's. Now George and I have come to share a time of life concern. Both of us feel compelled to do what it takes to assure that what we know and the body of work we've built up, he with his boot shop and I with our farming journal, have the best chance of continuing on beyond us. We feel the need to share all of our secrets with those who might want to get a head start in valuable hands-on work. Maybe it's an archaic notion, maybe when today's young adults get to our ages they won't share this compunction. That would be disastrous for humanity because it would spell a complete end to tradition. George and I sense that we are feeling an instinct to protect our traditions by preparing for the handoff.

I'm the one who will stick his gnarly neck out and say that within this instinctual urge to hand off our mantel is, ironically, a basic societal need for self preservation and self improvement. Society desperately needs us to succeed with these last chapters.

Up On The Roof

It was 1960. I was a thirteen year old city kid. Seven of us in the family drove down in our '55 Buick to a cousin's beach house on the western side of Baja California. That's in Mexico. Whatever the reason, I was told I had to sleep on the roof, away from everyone else. I remember a flat-topped building with vegetables and herbs growing on the dirt roof of this large rough structure. It was surrounded by grass-capped sand dunes except for the beach front. I remember smells, the surf sounds, and a blanket of stars so close I was sure some had snuck into my clothing. Though it was clear I was being punished and separated, I was thrilled to spend those few nights like a bony swizzle stick in a fabulous sensorial cocktail.

Our cousins were Hispanic with no English. I was forbidden, from the age of five years old, to speak Spanish. As Spanish was my first language, the immersion those few days in rolling, folded chocolate word sounds was like a blanket. I was jealous to witness the happy comfort of my mother as she lobbed Castilian phrases into the waiting air of that short visit, jealous that we never saw that "belonging" comfort in our own home.

Walking on the beach I was fascinated to watch saddle horses being ridden in the surf's shallow back wash. Must have been obvious because a tall Cinnamon brown gelding approached me with what looked like a ten year old boy on his back. After he figured out we had to speak English for my enjoined understanding, he asked if I wanted to ride. He said he would let me ride his horse for a quarter. I told him I didn't know how. With innocence instead of courage I slid up on the bare back of the tall horse. A rope came round the neck and withers fastened to either side of a braided cord halter. No bit, no hackamore, just a halter. The grinning boy explained by motions how to pull the horse's head in the direction you wanted to turn. He slapped the animal's rump and we walked off along the edge of the receding surf.

Looked over my shoulder and the boy was gone. Didn't matter. the comfort I felt was real, comfort not security. The horse ambled for a while then entered a bouncing trot. I gripped mane and rope. He made a slow wide arch and trotted away from the ocean and towards the dunes. Then he broke into a long lope. Now I was frightened. I leaned forward, head to one side, and hugged his neck in an effort to stay on. We hit the grassy cap of a dune and he dug in and leapt forward. We were airborne for a very long two seconds and as I looked down I saw two terrified people, a man and a woman laying on the sand in an embrace, shielding their heads from flying sand as they looked up at the underbelly of some great beast its form distorted by the sun directly above and beyond.

We landed hard yet he still picked up speed then slowed as we approached buildings. There at the open gate stood the little Mexican boy flipping my quarter and whistling for my borrowed charger, my Bucephalus, my first horse. The horse stopped, I slid off, and the boy followed his steed back into the farmyard. I walked back to the beach house weighing my adrenalin rush.

The story of my first experience with a horse is very much a curious empty box. It actually forebode of nothing yet it has always been there, sock-like, to take on whatever relevance or meaning I needed to apply. For me sleeping at the edge of the sea in a rooftop vegetable garden and feeling the immersion in a strange yet intensely familiar language is all very important to the story of that first ride. So as I think about beginnings and all those things that go into shaping a life's work I find myself pulling on the sock that is this story as I try to understand the force of my earliest attractions to farming, art and horses.

People ask this question in many shifting forms; "What made you choose the life of an old-fashioned horse-farmer?" Sometimes I answer it sometimes I don't even make an effort. But now I'm intensely interested in understanding how we must crack the new armors of the young if we are to "get through to" candidate novices, those people who think they want to do what we do. And we must get through to them if we are to complete the hand-off, the passing of the proverbial baton. So that means we have to honor all these questions and make good attempt to honestly and completely answer them. To that end, and pointing back to that story of my first horseback ride, I tell that I had experiences which instantly cemented my attraction to horses and farming and the creative arts. Those stories have become more important than any polemic about the politics of these life choices. People are drawn to the evidence that our most human of attractions might be shared. Later, perhaps much later, we can get around to how to do these things. In the beginning, we need to allow the sharing to flourish and the attractions to be gathered as definition of our future.

Union with the Work

The attractions, realized, understood or catalogued, though important for motivation don't necessarily provide a foundation for learning. That requires building with some foresight. When you build, helter skelter, with little or no thought for the long haul, you invite that what you build will be less than it might be. Your start must be about building a foundation. At whatever age or background you might hold or have, if you decide to embrace a craft, learn it, practice it hopefully some day master it, how you start, be it deliberate or happenstance, - how you start will affect your success. When I started out, in the early seventies, to learn the craft of working horses, I had absolutely no sense of this. And there was no cultural framework to try to instill this in me. Way back then society was well on its way in the wholesale process of demeaning, belittling, devaluing the life of the craftsman especially within a discipline like agriculture where the big money had settled on industrial process. Back then there were no schools, no books, no videos, no computer internet elements, nothing that said come here to learn. I know now that I was lucky beyond measure for the genuine relationships I stumbled onto that gave me internship with true masters. One in particular, Ray Drongesen, gave me the foundation necessary to allow that what I was to build would carry for a lifetime, and perhaps beyond. Ray taught me to get the mechanical stuff right, to pay attention to details and to quit trying to analyze everything. He taught me to hurry up and get it done. Before any ad agency sucked the life out of the words he repeatedly told me to '*just do it*'.

Mentors and mentorship come often as casual and fleeting bits of experience, not always structured and prescribed. But the structured internship can be most durable. I do, or try to do, many things with my days including farming and painting and writing. I never saw myself as one who would become a writer. While attending the San Francisco Art Institute in the sixties I was required to take English. Didn't do well at it at all until a visiting instructor changed everything for me. He required all of us art students to sit on the floor cross-legged and just listen as he told stories. "Now," he said, "tell us your own stories and hear yourself writing those words down." Every so often he'd interrupt and say "cool" or say "listen to yourself, you don't talk like that." His name was Ken Kesey and his subsequent literary contributions are legendary. As a teacher and as the one who I consider to be my writing mentor he was a liberator. He allowed me to give up on notions of what it meant to write correctly and embrace what it meant to communicate, to tell a good story and to see in that story its value out away from the teller. To see the story as a hollow conveyance. A one sock.

Though less than perhaps at any time in modern history, mastery of a craft, of a handmade skill-set, of a human way of working, of an art form is still valued by society. To be a masterful farmer, or musician, or carpenter, or writer, or ... is, by consequential effect, to be a positive contributor to the future of all humanity. And it is an accomplishment to be earned over a lifetime of devotion to the requirements of that craft; routines, rituals, menus, formulas, procedures and ever-unfolding spectrums of possibilities for adventure and innovation - it all goes together to make of that farmer/artist a seamless union with the work.

Some disciplines offer devices and opportunities to see a life's work, a mastery, passed forward in time to new generations. It comes when there are devices to record and store the evidence and the recipes. Books, artifacts that reveal themselves completely, sometimes film, sometimes recorded sound. Even so, to study a set of original paintings by Rembrandt or Georgio Morandi or Albert Bierstadt or Winslow Homer or Charles Burchfield - masters all - may reveal to the thirsty and prepared eye many secrets of process but it never will reveal the realities of natural facility, context, reflex, failings, urgings, and captured hungers. If these things, specific to these artistries, are to be learned completely it must come from shared experience and/or from a hard long road of parallel experience.

Sir Albert Howard was a masterful soil scientist, farmer, and lay nutritionist. Today some of us credit him with the beginnings of "Organic" agriculture, while others of us see him as an ordained protector of the biological world. Howard, thankfully, wrote books and papers. (See *An Agricultural Testament* for one.) He established agricultural test stations, and he taught others. They in turn, along with his written words, allowed that his exceptional and incredibly valuable work come forward to us. But Howard did not spring from a vacuum, he came to us courtesy of the influences of other powerful thinkers and doers. The list of names goes back in time just as those who were influenced by Howard directly have moved forward in time. No brainer you might say. You recognize the value in these handoffs? Then why is it we find ourselves in these dark days falling so far back? How is it that 'apprenticeship' has come to mean so little to the individual and to society at large? How is it that tradition is now viewed as silly?

Stop The Bus, This Is Where I Get Off

Old age is supposed to bring wisdom, if that is so then I know by fleeting experience that in my case I was wise for about twenty minutes one Wednesday afternoon about three years ago. Since then I've been struggling to keep the boat afloat and wisdom often got in the way.

twenty two

saving for what?

"Will you listen to yourself?"

The first thing you notice upon meeting him is that voice, slow and oddly musical, immediately recognizable… "It's something I worked at," he said. "I grew up listening to my favorite actors, and they all had unique voices. James Stewart, Humphrey Bogart, James Cagney. And when I started acting, I did not have a good voice, so I had to actively experiment with it and see if I could find rhythms in it, or break it up, or mess with it in some way."
                       - speaking of the actor Nicholas Cage

"I am that false character who follows the name around."
                - Don Delillo paraphrased

    Reading the words of Nicholas Cage I am reminded that I began my life's work in farming and art wishing I looked, sounded and moved differently. I wanted to have the character of a character. I had this secret wish to be one of those old sidekick types, like Smiley Burnette, or Raymond Hatton, or Gabby Hayes… only I wanted to be my own sidekick - not second fiddle to someone else who was in charge. There was this notion that

I might dress up my own life with humor and harmonica-ed tap dancing. And when things got serious the attention would just naturally slide sideways to the grownups in the room. I wanted that because I saw no future for me with the serious stuff. I wanted that because to my way of thinking the grown-ups in the room were off the mark. They could have the controls to the ship. Me and the girls in the room knew who was really in charge.

That fixation I had on the sidekick business always seemed to naturally parallel my deep interest in farming and art. I saw farmers as grizzled old Barry Fitzgerald characters with a kind touch for the leaf of a cultivated plant and a twinkling nod for the others sharing the space. And I preferred those artists, like Alexander Calder, who seemed to embody a jovial grampsterism. I wanted to be like them.

But in my middle years I thought all of this silliness from my youth was just that, silliness. And I grew to understand, or so I thought, that we need to accept and embrace who we are at our core. That you can't change your voice, or how you stand, like some paid actor. Well to add contradiction and paradox to the mix, now I know for certain that I don't know anything for certain. Yet, I do have notions and perceptions that I can't seem to shake. One is that how we see ourselves when we are young is a powerful force in our lives. A force that can set us up for success or failure or a mess of regret. Makes me wonder, these days, how young people might see themselves. Who might they want to emulate?

I'm an old man now. I feel the corrosive constant crawl of evaporating time - but I suspect I feel it less than many folks because I have my uplifting hunger for the next bits my farming will bring me. At many ages folks are heard to utter the deadening words "there is nothing to look forward to". The words seem to come from tragic personal circumstance, from the regrets of a life spent waiting, and from the vacuum of a livelihood without life.

Two young brothers visited our ranch recently to discuss a partnership that would have them adding fifty bee hives to our small cluster. We talked about the best locations, the shared-chore responsibilities, the challenges. On the surface it may have seemed to an outside observer to be a pedantic, even boring conversation, but to us it was exciting. It spoke to the vitality of what each of us knows, from experience. It spoke to "looking forward to…".

Early spring, visiting farms to look at possible consignments for our (former) annual auction event I was struck by the stories I picked up from the very different farmyards I viewed. Three in particular seemed to touch my reverie.

One was of an old ramshackle place that had been abandoned last fall. All machinery and debris had been stripped from the place save for the old leaning livestock buildings. The season had not yet allowed new plant growth so the view was of accumulated drying piles of cattle, sheep, chicken and horse manure. To an old farmer the manure was all good. I saw in this neat if crumbling panorama certain fertility, suggested history, and a kind of situational pregnancy. This place, in my admittedly slanted perspective, seemed to beg for a new young family of farmers to slip in leading a milk cow and carrying a box of layer chicks.

The next farmstead had been meticulously tended like a combination Japanese Zen garden and a western historical museum with rusted pieces of farming, ranching, mining, and ghost town memorabilia painstakingly presented in raked, raised-bed platforms. Though the history was on display it all seemed borrowed. No cobwebs, or misplaced minutiae, every single particle carefully in its place. Though it was fascinating to look at, there was no implicit invitation to engage the spot.

The third farmyard warned me from the first glimpse that to enter was to give up something. Young trees grew up through abandoned implements and garbage. Scattered vehicles sank into the ground from years of neglect. Old horses and cows lent rib shadows to the confused patterns of bits of blown-apart roofing and reclining rusted refrigerators. And the haggard single-wide mobile home seemed to be an argument with itself about utility. Then came the crowning contradiction; the golden people in residence. The old couple were in their eighties. She never got up from her recliner but bubbled with excitement around our visit. There was a promise in her smile that seemed to say "we've got things to show you." You could tell by the length of his cane that he had once been taller but the captive reflections in his eyes, marking the center line between his downturned handlebar mustache and upturned fedora brim had him fill all available space in a small room holding four people.

Within two sentences it was established that there was no need to waste a second, we had many things in common to bring into useful reflection. Any first impression sense of tragedy, from the yard view, was gone replaced by the vital rapid bits of over-lapping tales. We spoke of pack animals, laying hens, old springing cows, saw milling, rigging John Deere model Ls for garden work, starting a Paint team, leatherwork, vaquero traditions, western painting, book writing, Colorado, Juniper lumber, mounted Longhorn steer heads. And without an ounce of regret he said, rising up several inches, "I got so many things yet to do that I couldn't do 'em all if'n I lived to 200. Seen 83 winters so far." This magical man, his majesty stripped from him by his infirmity and by the abuse of his young relatives, was there in front of me (and himself) to make a case for how his future still had the power to positively flavor not only his remaining days but any assessment of his true place in life.

The first place was just that - a place - but one which spoke of possibilities.

The second farmstead was like some sort of storage, where valued memories were benched.

The third homestead was a backdrop in flux and of little consequence to the real view which was of a particularly fine humanity.

It speaks volumes that I was most drawn to the third farmstead. The people there suited me and gave me a dose of purposeful gratitude. The other two places had their attractions as well.

Musing about them I kept hearing the voices of other people saying "but we need some guarantee that this is going to work, going to make us some money. We can't afford to take a chance." And why not? I heard myself asking. Is it that threatening and dangerous when you go for what you are drawn to, what seems to give you promise and right identity?

You don't want to hear the old saw from me; the one that includes 'I started with nothing and had to work hard all the way'. But what you need to hear, especially if you are feeling hesitant about following your heart to a life working the land, is that I am an old man who took every chance to farm, followed the thread of my instincts, had a whale of a lot of setbacks and I would not trade that time so spent for any other life. I haven't made much money, I worked without medical insurance or any kind of safety net, and for these last five years I volunteered my time to keep a farming Journal going. When I would sell a painting or two or some of my books, that all went into the publishing venture and into my farming. No complaints, it has given me a golden life. So I say jump in there right now. Take the plunge. Yeah, you might fail, and maybe even several times. But so what? Failing at a full throated life is no failure at all because you'll be doing what you want to do.

All my life I have heard people talk about saving up for the life of their dreams. That usually entails charting a course and figuring a budget. Those two nebulous things - course and budget - they are the ultimate moving targets. When you worry yourself through that process you never have enough money, you never feel reassured. I have to ask, 'Saving for What?'

Others will anger at these words, pointing out that it is irresponsible to encourage people in general to take life-changing risks. And that at the core it is also anti-social because society is supposed to be the balance beam of humanity and to counter the prevailing notions of success and livelihood threatens society. To which I say society right now, as we know it, stinks. It is the result of our collective appetites, requirements and appointments buried deep beneath artificiality. Society is a nasty swirl of class conflict, collective impatience and pettiness. I say go ahead and threaten society with bold moves towards purposeful livelihood and creative fertility. Society will be the better for it, and so will you.

In my mind's eye I see an old photo of three people; the frightened young man wore faded blue jeans stiff from drying on the line and presenting the crisp pleat of a woman's hot iron, the old man's felt hat rocked back to allow his face forward to be first to catch everything in his path, and the woman her hands rolled into her apron front said "hurry up, I've got things in the oven". I see the photo and I wonder why I'm showing this to myself, what's it supposed to mean? Up pops this notion in my head that what I'm looking at are three different manifestations of vitality. The young man guessing himself forward. The old man face to the next wind. And the old woman always setting the next table, already forward.

Our vitality is so incredibly important. Yours, mine. We should be protecting our vitality? It's only natural for young folks to worry some about the choices they face, natural and intelligent. But somehow we need to help them to see that this modern world is confusing their options not helping them. For those of us at these later stages of life it is time to have the courage to make every case against how so much of industrial chemistry, genetic engineering and information technology race to extinguish humanity. We need to turn the lights on so that it's obvious how most of the highest paid, most lucrative positions in modern society are toxic to everyone. There is that deadly irony of how it is that a few people are being paid enormous sums of money to orchestrate the end of our species through supplanting human effort.

There may be some calling equal to farming but none superior. Saving for what?

twenty three

old man farming

With the passage of our collective time, through the corrosive sieve of plastic disconnect and modern indifference, who will survive to remember; remember what was long ago inherited, what was long ago desired, what was long ago left to us to protect, all that experience which taught us those difficult lessons of how to fit within ourselves? And how to know who we are, what we might choose as our very own embraceable livelong endeavors. To be of an age when such questions come with urgency is ...

My father taught me how to work, my mother tried to teach me to avoid sentimentality. My father succeeded, my mother never came close. To the contrary I have found strength in the re-imagination of new forms of sentimentality, new that is until I discovered that they were the old original definitions and forms. For sentiment of old, along with romance of old, embraced and grew from conscious understanding of structure and history. Once upon a time there were two ruling observations to the social history of mankind; naiveté and sentiment.

It was a pasty, mustard-yellow, half-ton '80 Chevrolet pickup truck the entire body of which was riddled by small dents. I bought it at auction years ago. It belonged to an old rancher whose family fondly referred to as Mr. Magoo. In his later years he could not see well. He only drove at home on the ranch, and he drove this truck. He'd drive slow until he bumped into something, then he'd back up a little and turn one way or the other and try again. In this fashion he 'felt' his way around the ranch. And in this fashion he dented up old yeller. The surface of the vehicle was like a reverse brail, a record of 'felt', as if Mr. Magoo used 'old yeller' like a big motorized blind man's walking stick, feeling out around him as he moved through his farming world. The pickup truck was in great shape, internally. It had low miles. But externally it looked like a real disaster. I bought it cheap and drove it home to become our ranch "crummy".

Today I found myself remembering that old truck as my current physical condition seemed to parallel that of Mr. Magoo's Chevrolet. I too am badly dented with dozens of ridiculous problems from failing eyesight, to loss of hearing, to sundry nerve damage problems, to dizziness, etc etc. The parallel continues as I too have bumped around, feeling my way through life seldom able to see what has been right in front of me, better perhaps at the long views. Even so, deep inside I am in great shape - way deep inside. Happy to be where I am and farming still. Sure, I'd love to get the use of my hands back so I could drive my horses again and return to my fountain pens and paint brushes - maybe that will come. But my story, my condition nearing on seven decades of life, is nothing unique and only worth mentioning to give some authority to the following short treatise on the comic, heroic, socially invaluable tragedy of old men farming.

There is a giant old Juniper tree in our front yard which, twenty-six years ago, provided luxuriant shade and bird habitat. As it slowly dies off the shade aspect becomes goofy, useless and spotty. But oh the posture! Bare limbs akimbo this beast of a tree twists out and up, only thing is it has begun to lean - towards the power line! I know that some time soon I may have to take remedial action but for now I look upon the old tree and ask myself 'what is the lesson here?' Should I be so quick to judge, to act? After all, I too am leaning towards some proverbial power line. How long can both of us hold this pose? And what might be the value of this?

We had a twin to this tree which for many years stood over my blacksmith shop. One day, in a stiff wind, I watched and listened as it sucked its huge roots out of the ground and slowly fell on to that shop roof. Another old man, dear friend Jean Christophe of France, chainsaw in hand, climbed high up in that monstrous tree, leaning at 45 degrees up over the building, and cut pieces of it off from the top, lowering by cord each chunk to waiting Natalie on the ground. It was like some utilitarian circus act played out for an audience of 25 peafowl, saddened because this had been their favorite nighttime perch, and an old farmer who wished things like that ancient tree could have stayed on forever.

What is notable with this writing and these observations, is that today I put far more enthusiasm and energy into the notion of planting more trees, many more. I don't think or worry so much about the aged frail trees now leaning. There in lies part of the secret strength of old men (and women) farming. We reach back and forth in time with the earned intimacy of overlapping growth cycles and how we might magically influence those. We know, completely and well, where farming choices fit in the superb uncertainty of the natural world. And we, if of those long suited to the venture, embrace the odds.

In our front yard stand two massive tall poplar trees which I planted 25 years ago. (If you saw them you might think they are three times that old.) I know by their example that, with good fortune and planning I might plant more which will have that sort of dramatic effect 25 years hence. No guarantees of course, but should I succeed - what magnificence!

Along with big old Ponderosa pines, our ramshackle hundred year old western ranch is home to many ancient scraggled and gnarled Juniper remains which 'act' as place marks, as headstones, nature's allowed and preferred notations. Occasionally we humans moving through life stumble on to embraceable endeavors which allow that there might be some evidence of our lifelong efforts to linger after our passage. Maybe one of those endeavors for us is the planting of additional trees. I like the odds.

Tree planting seems a quart-sized analogy for the craft of farming. Perhaps its the obvious chance of durability, and then they are those pesky vagaries. Who can predict, who can know, how a series of lightning storms, heavy wind, dry weather, heavy snowfall, bouncing temperatures and sideways moisture, might yank at roots, split limbs, invite pest and pestilence;

who can know? Yet we old farmers remain enchanted by variation.

This year we divided one field into six 7 acre lands. Careful to record each variation; we planted four successive lands to an Oat and field Pea mix for hay and soil health. Each land was tilled differently and the four were planted over a period of 5 weeks to stagger harvest times. The fifth land was planted to Buckwheat intended to be a cover crop and soil envigorant. The sixth land was left fallow. We irrigated the planted fields and kept a record of performance. As of this writing we are nearing harvest of the first land. The growth of the five lands has been dramatic and dynamic in each moment, but for us what has been equally felt have been the floating questions about what crops will follow in each land as we progress with this multi-year rotation plan. Will we plant oats as a nurse crop with legumes and grasses after tilling in the nitrogen-rich Buckwheat? And will we do the same on the fallow land so that we might be able in the following spring to gauge the advantage of the Buckwheat? Will we follow Oats and Peas with Wheat or Barley, or experiment with Chick peas? Will we allow one land as a strict planting of Timothy? Will we plant cereal Rye in these irrigated plots or save it for the dry-land?

For forty-seven years I have practiced what some people choose to call organic farming. We use no chemical fertilizers, pesticides, herbicides, fungicides or store-bought poisons in our farming, but that is not how we choose to define ourselves. We don't want to be known for what we don't do. We want to be known for what we choose to do. We prefer to think of ourselves as farmers. Over the years the "official" organic approach to agriculture has become institutionalized to an unhealthy degree. Today it is a monetized, highly structured, policed private club. There are rules, there are fees, and there are rituals. If you do not follow them you will find yourself ostracized or afoul of the law. If you label your products as 'organic' without formal federal certification you are a criminal. We are not certified and have no intention of ever applying to that process.

A young farming zealot recently made a curious set of observations about me. She said by way of questions "What do you mean you're not certified? I know you're an organic farmer. You know it's the better way to go but how can you call yourself organic if your not certified? I mean let's face it, it doesn't matter how you farm if you aren't certified, because you can't reasonably prove to anyone that you farm organically unless you allow neutral federal testers to check you out regularly. If the federal government didn't offer a certification program anyone could claim they were organic and the word would mean nothing. You don't want that do you? I mean, honestly, how do you see yourself? What do you call yourself?"

I answered, "old man farming".

Once was a time I would have taken her on in two-fisted argument, but not anymore. I'm in a new zone, old man farming.

Anyway, I'm too busy to mess with such nonsense. Instead of age putting the damper on our future farm plans it's lit new flames. We have things we are doing now on our place that won't come to fruition for three to five years and that suits me to a T. And as for any membership, compliance, or government control - it don't apply in my zone. There is that terrible cliche about old folks - they say of us "why he's so old he has nothing to lose". Truth be known we are so old we have everything to gain and that much more to contribute. I've forgotten where I learned it - perhaps from my father, to win a race or complete an arduous task the secret is to see the goal line far beyond where it actually is. That way you'll fly by it. As I approach my seventh decade I imagine projects that will take three decades to complete. So whenever my life is done I will be far beyond that point in time.

The meaning of sad, the meaning of hopeful. The old juniper crag.

Late life musings certainly take first impetus from the work at hand, much more than from memories of what is past. Of course that changes if we are bed or chair bound and out of the work-a-day cycle. In those cases the tableau stretches back across our spent time. Spent, spending, built, building, oiled and oily.

Occasionally there are specific memories which surprise us with durability and application. As I find my old self working harder than ever at farm and ranching tasks, bumping along like Mr. Magoo in his yellow pickup, my brain reminds me that I gave that truck to a ranch-hand of ours, Jessie.

You've heard the term 'he had a screw loose'? Jessie's entire brain was one magnificent loose screw. Jessie had 'outlookitis', he'd look out at the world and it would carry him off to flights of bizarre conclusion and self assessment. He would leap like a ballet dancer from unrelated point to scattered premise all the while oblivious to how he came off and thrilled to be in control of the solution to all matters practical.

My friend Tygh Redfield, the man who introduced Jessie to us and who holds title as Jessie's existential biographer, reports that once upon finding a mess of empty beer cans in his hayfield he approached then neighbor Jessie for an explanation. With a raspy inhalation he said, "Gol' Tygh, last night I was playing catch with the moon and I guess the moon won."

Tygh somehow convinced us, years ago, that Jessie ought to come live on our ranch and be our handyman. He did that for a year. Even though I have fond memories of the man I am surprised still that I did not chase him off early on.

But he was so uniquely endearing. And that came in large part from his way with certainty and corrective respect. For example, he thought it was a sad mistake that my folks would give a girl's name to me. He couldn't bring himself to call me Lynn, instead he'd always call me Glen but with a happy preface, he'd shorten Golly to Gol' and address me with the setup as "Gol' Glen!" always followed by an odd observation. "Gol' Glen, this here is like some sort of real old-time cattle ranch, with cattle and horses and trees and everything! Gol' you got to love it Glen! There ain't no more places like this. And you and me we're just like a couple of old cowboys just cowboying and stuff. Gol'!"

Made me smile and park my mature self so I could take out my compassion for the old guy, except...

One morning, as I had requested, Jessie took fencing tools and drove the half mile of county road repairing our five strand barb wire fence. Mule deer damage. Three hours later as I left to drive to town, I saw Jessie and his little pickup on the other side of that same fence driving along at two miles an hour. He stopped, got out, and tapped at a post with his fence pliers while waving to me. "Jessie", I asked "what are you doing? I watched you earlier repairing this fence?" With rasping inhalation he answered "Gol' Glen, you gotta fix these old fences from both sides."

Jessie fell completely in love with 'old yeller' that dented pickup truck, and the back-story of Mr. Magoo the rancher. To him here was evidence that all was right with the world. "Gol' Glen, do you figure you could ever sell me that old truck? She and I would do good together, I just know it." There was something insistent in that implied logic, that this man and vehicle belonged together, so I sold it to him cheap. Probably should of held on to it, but there's that sentiment-business governing ways forward. Jessie's long gone now likely in charge of meaningless limericks up in heaven. He only factored in as a piece of one year in our long history, as a couple of small dents in our trajectory, feeling our way along.

But each of these specific little storied pieces of living rhythm played out as layers against the backdrop of trees, and the planning and planting of trees, fields and crops, livestock, and facilities - they give this old man a real charge forward.

Chronology offers us a peak at the comic aspects of relative age. I have dear friends in their seventies, eighties and beyond who, upon learning I am 67, pshaw their ways toward reflecting on all the things they coulda/shoulda done when they were as YOUNG as I am now. And that invariably brings us to discussions about odometer readings and the relationships between roads travelled and miles accumulated. How do I explain to the uninitiated amongst them that my accumulations reach forward and backward in time putting me squarely in the old man farming zone, where the numbers mean nothing?

Praying to you my fellow travelers, don't take from me my comforts, don't strip from me my stripes, allow me my delusions that I might continue to laugh at my own jokes and hurry at my own pace. For mine is the power of the old cuddler of connivance and your last order of business is to allow me a brief moment to intervene that together we might take some credit should the outcome satisfy.

"We ask the forgiveness of our sustained arrogance though we feel its furtherance in the asking. We move inside our gullibility afraid to ask of ourselves the surviving eye. We took the offerings of commerce and the false communion of idiots for hire. With an explosive expanding greed we forgot our children, and we left their children at the edge of the last conceit. How dare we now moan?

Ah, but the good news is that 'the great collapse' or the 'second great depression' gives us back our true songs, our handmade artistry, our blood-soaked words, and most valuable of all an opportunity to revisit enlightenment. We are all alone in this and the poets of connivance will win out. The greatest tool of a poetic connivance shall be a well-mounted sincerity; never the goal always the tool."

-from *"In Others Words"* by Lynn Miller

About the author: Lynn R. Miller was born in Kansas City in 1947. He is a painter, farmer, horseman and writer. He is the founder of Small Farmer's Journal and, for 38 years, has served as its publisher/editor. He has authored many books, articles and essays. He and his wife, Kristi Gilman-Miller, live on and work Singing Horse Ranch in Central Oregon.

To view images of Mr. Miller's artwork visit
www.lynnmiller-artworks.com

To purchase copies of his other book titles and/or for information about the International quarterly, Small Farmer's Journal, visit
www.smallfarmersjournal.com

Or write to: Small Farmer's Journal
PO Box 1627, Sisters, Oregon 97759

800-876-2893

Coming soon
by Lynn R. Miller

The Art of Working Horses

Talking Man (novel)

In Other Words (poetry - postings - drawings)

The Brown Dwarf (novel)

What 'm I Bid (stories)

Horsedrawn Drills & Planters

Also by Lynn R. Miller
and available from www.smallfarmersjournal.com

Farmer Pirates & Dancing Cows
Essays

Why Farm
Essays

Starting Your Farm
Guidebook

Workhorse Handbook Second Edition
an operator's manual

Training Workhorses/Training Teamsters
teaching

Horsedrawn Plows & plowing
the tool and the work

Horsedrawn Tillage Tools
comprehensive

Haying with Horses
an operational manual

The Mower Book
for horsedrawn

Thought Small
poetry and drawings

The Glass Horse
a novel

www.ingramcontent.com/pod-product-compliance
Lightning Source LLC
Chambersburg PA
CBHW071436080526
44587CB00014B/1864